SPARTAN CODE REFERENCE LIBRARY / 001

SPARTAN CODE: MATH

BASIC KNOWLEDGE SURVIVAL

MEASURE / MODEL / CALCULATE / VERIFY

FRANK CASTLE

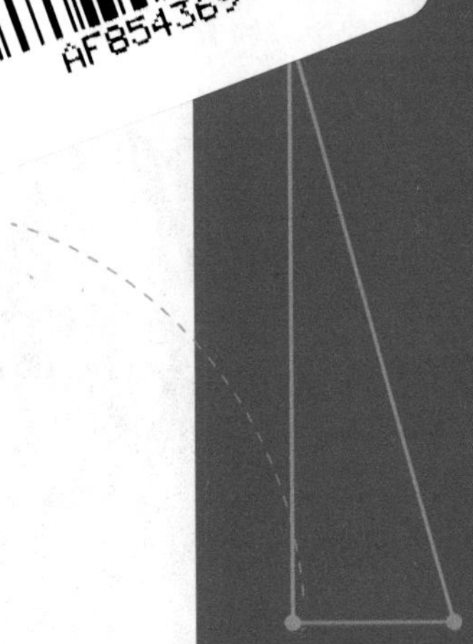

Paperback Black-and-White Edition / 4DAT / ISBN 978-1-972327-18-0

SPARTAN CODE: MATH

Basic Knowledge Survival

Paperback Black-and-White Edition

Original work: *Elementary Practical Mathematics* by Frank Castle, first published by Macmillan and Company, Limited, London, 1899. The base edition used here was digitized from the University of Michigan copy and is identified by Google Books volume Abk2AAAAMAAJ. The original work is in the public domain.

Source transcriptions are corrected cautiously and cited by the printed pagination of the 1899 edition. Mathematical corrections and modern extensions are identified rather than silently attributed to Castle.

Published by 4DAT LLC Panama City, Florida, USA

Paperback ISBN: 978-1-972327-18-0 First edition, 2026

Publisher website: https://www.iamfrankcastle.com

Build date: August 9, 2026
Edition source: modular XeLaTeX, 6 x 9 inches.

Why Rebuild This Book?

Frank Castle wrote for artisans and technical students who could not see how classroom arithmetic, algebra, and geometry connected to the shop or factory. His answer was practical mathematics: begin with a measurable situation, form a mathematical representation, calculate, and verify.

That sequence still works. The tools have changed. A steel rule may now share the bench with a laser measure. Logarithm tables have become library calls. Squared paper has become a plotting window, and the planimeter has become numerical integration. The intellectual habits beneath those tools remain: units, scale, approximation, structure, and a willingness to check a model against reality.

This edition therefore rebuilds rather than merely reproduces. Castle's original sequence and page references anchor every chapter. The prose and notation are modern. Worked examples use current units and situations. Computational extensions show what the same mathematics becomes when placed inside a spreadsheet, a short program, a sensor, or an interactive model.

The editorial boundary matters. Public-domain source material is labeled as the 1899 idea. New explanation is labeled as modern. Extrapolation is labeled as new. The reader should always know which century is speaking.

The result is meant to be used with a pencil, a measuring tool, and whatever computational aid fits the problem. Mathematics becomes durable when the reader makes it do work.

How to Use This Book

Each chapter repeats a seven-part rhythm.

1. **Original scope** identifies the corresponding 1899 pages.
2. **The 1899 idea** gives a short quotation or faithful summary.
3. **Modern explanation** updates notation, terminology, and units.
4. **Worked example** carries one problem from data to checked result.
5. **Push the idea** connects the topic to computation, engineering, or uncertainty.
6. **Companion lab** specifies an experiment that can be run with a calculator, spreadsheet, short program, or measuring tool.
7. **Check yourself** supplies exercises; answers are collected at the back.

Do not read the worked examples passively. Cover the next line and predict it. Estimate the scale of the answer before calculating. Keep units attached. Use each lab to vary one input at a time, then explain the observed change in words. A correct number without a model is fragile; a model that survives variation is understanding.

IN TODAY'S NOTATION

Throughout the book, *exact* means exact within the mathematical model. Measured data still carry uncertainty. A numerical display with many digits does not create precision that the measurement never possessed.

Contents

PART I

Number as a Tool

CHAPTER 1

Measure What Matters

Lab 01: Measurement Bench

Original scope: Chapter I, printed pages 1–6

Length, units, standards, conversion, resolution, and the discipline of attaching a unit to every measured number.

THE 1899 IDEA

Castle defines measurement as comparison with a selected unit. The numerical value tells how many times that unit is contained in the quantity. He treats length, mass, and time as the foundational quantities from which the practical student builds everything else. His comparison of the yard and metre is also a lesson in system design: decimal relationships reduce clerical work and error.

Source: Castle (1899), pp. 1–6.

1.1 A number is not yet a measurement

The statement "the plate is 240" is unfinished. It might mean millimetres, centimetres, inches, grams, or degrees. A measurement has three pieces:

$$Q = n\,u, \tag{1.1}$$

where Q is the quantity, n is the numerical value, and u is the unit. Writing 240 mm preserves the meaning. Writing only 240 throws away the comparison that created the number.

Modern SI definitions are tied to physical constants rather than particular metal bars. The metre is defined through the speed of light, and the second through a transition frequency of caesium. The practical lesson is the same as Castle's: a standard must be reproducible, stable, and shared.

HISTORY THREAD: Mathematics has many workshops

In ancient Egypt, reliable measures supported fields, building, taxation, and trade; a royal cubit turned the variable length of a human forearm into a shared standard. Mesopotamian scribes likewise developed tables and procedures for practical calculation. These records remind us that mathematics did not begin only when someone proved a theorem. It also began when communities made quantity public: a length another worker could reproduce, an account another scribe could audit, a rule another generation could teach.
The philosophical shift is small but profound. Measurement changes a private impression into a claim that can be checked. In that sense, the first standard rods were not just tools. They were agreements about reality.

Historical note: MacTutor, "The history of measurement" and "Egyptian mathematics".

1.2 Conversion is multiplication by one

The equality 1 in=25.4 mm gives the dimensionless factor

$$\frac{25.4\,\text{mm}}{1\,\text{in}} = 1. \tag{1.2}$$

Multiplying by this form of one changes the unit without changing the length. Arrange the factor so the unwanted unit cancels:

$$7.5\,\text{in}\left(\frac{25.4\,\text{mm}}{1\,\text{in}}\right) = 190.5\,\text{mm}. \tag{1.3}$$

This cancellation method is safer than memorizing whether to multiply or divide. The units themselves tell you.

WORKED EXAMPLE: Stock allowance

A finished shaft must be 18.00 in long. The saw operation needs 6 mm of total trimming allowance. What minimum stock length should be cut in millimetres?

$$L_{\text{finished}} = 18.00(25.4) = 457.2\,\text{mm},$$
$$L_{\text{stock}} = 457.2 + 6 = 463.2\,\text{mm}.$$

The input 18.00 in is stated to the nearest hundredth of an inch, so reporting micrometres would claim unjustified precision. A practical cut instruction is 463 mm minimum.

WORKED EXAMPLE: LiDAR scan to metric order

A phone LiDAR scan reports a wall run as 11.35 ft. Imported flooring is sold by the metre. Chain two factors of one so the unwanted unit cancels at each step:

$$L = 11.35(12) = 136.2\,\text{in};$$
$$L = 136.2(25.4) = 3459.5\,\text{mm}.$$

The app's step of 0.01 ft is about 3 mm, so the centimetre is the last defensible digit: order 3.46 m of run.

1.3 Resolution, precision, and accuracy

A ruler marked every millimetre has a resolution of 1 mm. Repeated readings that agree closely are precise. Readings close to the true value are accurate. A mis-zeroed calliper can be precise and inaccurate at the same time. These distinctions explain why extra digits on a display are not automatically extra knowledge.

PUSH THE IDEA

Digital sensors turn physical quantities into counts. If a 12-bit converter covers 0 V to 10 V, one count represents approximately $10/4095 = 2.44$ mV. Calibration then maps counts to a physical unit. Castle's unit comparison has become a software function, but it remains a comparison against a standard.

COMPANION LAB: Measurement Bench

Choose a unit system, enter a measured value and instrument resolution, and convert between millimetres, inches, metres, and feet. The lab displays unit cancellation and rounds the output to a precision justified by the input. Test the same object with coarse and fine instrument resolutions and compare what changes and what does not.

CHECK YOURSELF

1. Convert 3.75 ft to millimetres.
2. A calliper reads 24.62 mm with resolution 0.02 mm. State a sensible measurement interval.
3. A 10-bit converter spans 0 V to 5 V. Find volts per count using 1023 intervals.
4. Explain how a measuring tool can be precise but inaccurate.

CHAPTER 2

From Length to Matter

Lab 02: Density Tank

Original scope: Chapter II, printed pages 7–21

Derived units for area and volume, density, relative density, displacement, and Archimedes' principle.

THE 1899 IDEA

Castle builds area and volume from length: the area unit is a square whose side is one length unit, and the volume unit is a cube whose edge is one length unit. He then connects volume to mass through density and uses displaced water to make the mathematics experimentally visible.

Source: Castle (1899), pp. 7–21.

2.1 Why conversion factors must be squared or cubed

Because 1 m=100 cm, a square metre contains $100 \times 100 = 10{,}000$ square centimetres:

$$1\,\text{m}^2 = (100\,\text{cm})^2 = 10{,}000\,\text{cm}^2. \tag{2.1}$$

For volume, the factor is cubed. This is why converting 1 m^3 by a factor of 100 gives a million, not a hundred, cubic centimetres. A unit conversion applied to length must be applied to every independent dimension.

2.2 Density binds mass to space

Density is mass per unit volume:

$$\rho = \frac{m}{V}. \tag{2.2}$$

Relative density, historically called specific gravity, is the ratio of a material's density to a reference density. For solids and liquids, water near 4 °C is commonly used. Because it is a ratio of like units, relative density is dimensionless.

IN TODAY'S NOTATION

Castle often reasons in terms of weight. Modern technical work distinguishes mass, measured in kilograms, from weight, a force measured in newtons. Near Earth's surface, $W = mg$, but the two concepts are not interchangeable.

WORKED EXAMPLE: Identify an unknown metal

A machined part has mass 612 g. Submerging it raises the water level from 350 mL to 428 mL. Since 1 mL=1 cm^3,

$$V = 428 - 350 = 78\,\text{cm}^3,$$
$$\rho = \frac{612}{78} = 7.85\,\text{g/cm}^3.$$

The result is consistent with steel. The conclusion is evidence, not proof: several alloys have similar density, and trapped air would bias the result.

WORKED EXAMPLE: Will the floor take the aquarium?

A tank 100 cm long, 40 cm wide, and 50 cm deep holds $V = 100(40)(50) = 200{,}000\,\text{cm}^3 = 200\,\text{L}$. At 1.00 g/cm^3 the water alone is 200 kg; with about 30 kg of glass, stand, and gravel, the footprint of $1.00 \times 0.40 = 0.40\,\text{m}^2$ carries

$$\frac{230}{0.40} = 575\,\text{kg/m}^2,$$

nearly three times a typical 200 kg/m^2 residential floor rating. Volume grows by three multiplied lengths; the floor only ever sees the two of the footprint.

2.3 Buoyancy as a balance of forces

Archimedes' principle states that the buoyant force on a submerged body equals the weight of the displaced fluid:

$$F_b = \rho_{\text{fluid}} g V_{\text{displaced}}. \tag{2.3}$$

An object floats when buoyancy can balance its weight before the entire object is submerged. For a floating object, the submerged fraction is approximately $\rho_{\text{object}}/\rho_{\text{fluid}}$.

PUSH THE IDEA

Density measurement becomes a quality-control signal. A cast component with unexpectedly low bulk density may contain voids; a battery cell whose apparent volume changes

may be swelling. Combining mass, geometry, temperature, and uncertainty turns a nineteenth-century displacement experiment into a modern diagnostic pipeline.

COMPANION LAB: Density Tank

Enter mass, initial water level, and final water level. The lab calculates volume, density, relative density, buoyant force, and likely material families. Toggle a trapped-air error to see how a small volume bias changes the inferred material.

CHECK YOURSELF

1. Convert 0.45 m^2 to square centimetres.
2. Find the density of 2.70 kg occupying 1.00 L.
3. What fraction of a block with density 720 kg/m^3 is submerged in fresh water?
4. Explain why a hollow steel ship can float although steel is denser than water.

CHAPTER 3

Time, Angle, and Rotation

Lab 03: Rotating Dial

Original scope: Chapter III, printed pages 22–38

Time standards, sexagesimal angles, radians, arc length, geometric construction, and angular rate.

THE 1899 IDEA

Castle places time and angle together because both use non-decimal subdivisions and both appear constantly in practical mechanics. He develops degrees, minutes, seconds, and radians, then treats construction with straightedge, compasses, scales, and protractors as a way to verify numerical relationships.

Source: Castle (1899), pp. 22–38.

3.1 Three ways to measure angle

A full revolution is 360° or 2π radians. Radian measure follows directly from circle geometry:

$$\theta = \frac{s}{r}, \tag{3.1}$$

where s is arc length and r is radius. The ratio has no unit, although "radian" remains a useful label. Conversion follows from $180^\circ = \pi$ rad.

$$37.5^\circ \left(\frac{\pi \text{ rad}}{180^\circ}\right) = 0.6545 \text{ rad}. \tag{3.2}$$

Radians are natural in calculus and motion because arc length is simply $s = r\theta$. Degrees remain convenient for drawing, navigation, and human communication. The correct choice depends on the operation.

The third working system is the mil, built for speed rather than elegance. A milliradian is $1/1000$ of a radian; military compasses and fire-control instruments round the $2\pi \times 1000 \approx 6283$ milliradians in a circle to 6400 mils so the circle divides cleanly: 1600 mils is a right angle, 800 mils is 45°, and 640 mils is exactly 36°. The mil inherits the radian's small-angle rule: because $\tan\theta \approx \theta$ for small angles, an object of size s at range d spans about $1000\, s/d$ mils. One mil is one metre of target at one kilometre of range.

WORKED EXAMPLE: Range a figure with a reticle

A standing figure about 1.8 m tall spans 2 mils between the dots of a spotting-scope or rifle reticle. Estimate the range:

$$d \approx \frac{1000\,s}{\text{mils}} = \frac{1000(1.8)}{2} = 900\,\text{m}.$$

At 3 mils the same figure stands near 600 m; at 1 mil, near 1800 m. The reticle turns the small-angle rule into a rangefinder with no battery to fail.

HISTORY THREAD: Sixty, twenty, and the shape of time

The Old Babylonian place-value system used base 60. Its long afterlife is on every clock face and in the division of angles into degrees, minutes, and seconds. Across the ocean, Maya mathematicians built a largely base-20 positional system, including a sign for zero, and adjusted one place to serve calendar calculation. Neither base was inevitable. Each fit habits of counting, astronomy, and administration.

This is a useful warning against mistaking familiarity for necessity. Decimal notation feels natural because we inherited and practiced it. A representation is a tool chosen by a culture; once chosen, it begins to shape which operations feel easy and which patterns become visible.

Historical note: MacTutor, "Babylonian numerals" and "Mayan mathematics".

3.2 Cycles, frequency, and angular speed

Frequency f counts cycles per second. Period T is time per cycle:

$$T = \frac{1}{f}, \qquad \omega = 2\pi f. \tag{3.3}$$

Angular speed ω is measured in radians per second. A shaft turning at N revolutions per minute has

$$\omega = \frac{2\pi N}{60}. \tag{3.4}$$

WORKED EXAMPLE: Belt speed from shaft speed

A pulley of radius 80 mm turns at 1450 rev/min. Convert the radius to 0.080 m, then

$$\omega = \frac{2\pi(1450)}{60} = 151.84\,\text{rad/s},$$
$$v = r\omega = 0.080(151.84) = 12.15\,\text{m/s}.$$

The circumference check gives the same answer: distance per revolution $2\pi r$ times revolutions per second.

3.3 Longitude is kept time

The Earth turns 360° in 24 hours: 15° per hour, one degree every four minutes. Local solar noon, the moment the sun stands highest, therefore sweeps westward around the planet like a clock hand. A navigator who carries the time of a reference meridian can convert the gap between local noon and carried noon directly into angle east or west. For two centuries the hard part was the carrying: building a clock that pitch, damp, and temperature could not corrupt.

WORKED EXAMPLE: Longitude from a clock

At local solar noon a ship's chronometer, set in Greenwich, reads 14:24. Local time lags Greenwich by 2 h 24 min = 2.4 h, so the ship lies west of Greenwich by

$$\lambda = (2.4\,\text{h})(15\,°/\text{h}) = 36.0°\ \text{west}.$$

A four-minute clock error is a full degree of position, about 111 km at the equator, which is why the chronometer rather than the telescope finally fixed longitude at sea.

HISTORY THREAD: The carpenter's sea clock

After a fleet wrecked on the Scilly Isles in 1707 partly for want of longitude, Britain's 1714 Longitude Act offered up to £20,000 for a practical method. John Harrison (1693–1776), a self-taught joiner, spent five decades building clocks the sea could not corrupt: bimetallic parts to cancel temperature, low-friction bearings, a balance indifferent to a ship's roll. His H4 watch crossed the Atlantic in 1761 losing seconds while the rival lunar-distance method demanded hours of calculation. GPS is the same idea industrialized: satellites broadcast atomic time, and a receiver turns tiny clock differences into position. Navigation has always been applied timekeeping.

Historical note: MacTutor and Royal Museums Greenwich, "John Harrison".

3.4 Time is also measured data

Elapsed time is the difference between two timestamps, but clocks have resolution, drift, and synchronization error. When measuring a short event, human reaction time may dominate. Repeating the event over many cycles and dividing the total time can reduce relative timing error.

PUSH THE IDEA

Encoders report shaft angle as discrete counts. With 2048 counts per revolution, one count is $2\pi/2048 \approx 0.003{,}07$ rad. Estimating speed from successive counts introduces a trade-off: short time windows respond quickly but are noisy; long windows are smoother but delayed. This is the beginning of digital signal processing.

COMPANION LAB: Rotating Dial

Control radius, revolutions per minute, and elapsed time. Watch angle in degrees and radians, arc length, tangential speed, and encoder counts update together. A sampling control demonstrates speed quantization at low rates.

CHECK YOURSELF

1. **Convert 225° to radians.**
2. **Find the arc length for $r = 0.40$ m and $\theta = 1.2$ rad.**
3. **Find angular speed for 900 rev/min.**
4. **Why does timing 20 pendulum cycles usually improve a period estimate?**

CHAPTER 4

Arithmetic That Carries Meaning

Lab 04: Fraction and Average Workbench

Original scope: Chapter IV, printed pages 39–58

Operations, factors, common multiples, fractions, decimals, and averages as practical representations.

THE 1899 IDEA

Castle reviews arithmetic not as a catalogue of school rules but as a working language. Fractions describe parts and ratios; decimals support calculation; greatest common factors and least common multiples simplify repeated patterns; and averages compress a collection into a representative value.

Source: Castle (1899), pp. 39–58.

4.1 Operations are claims about a situation

Addition combines like quantities. Subtraction compares or removes. Multiplication represents repeated groups or scaling. Division represents sharing or a rate. The symbols are compact, but the operation must still match the physical story. Adding 3 m to 4 s is not a measurement; dividing them creates the meaningful rate 0.75 m/s.

Operator precedence is a convention for reading a line of symbols: grouping, powers, multiplication and division, then addition and subtraction. When the intended grouping matters, parentheses are inexpensive insurance.

HISTORY THREAD: Liu Hui and the algorithm as explanation

The Chinese classic now called *The Nine Chapters on the Mathematical Art* organized practical problems in surveying, taxation, trade, and engineering. In the third century, Liu Hui wrote a commentary that did more than preserve recipes: he explained why procedures worked. Counting rods made numbers movable on a surface, supported calculations with positive and negative quantities, and enabled elimination methods recognizable in modern linear algebra.

There is a philosophy of workmanship here. An algorithm is not lesser mathematics because it begins with a problem. A dependable procedure captures reasoning so another person can repeat it. Explanation then turns inherited craft into inspectable knowledge.

Historical note: MacTutor, "Liu Hui" and "Nine Chapters".

4.2 Fractions and decimals are two views

The fraction 3/8 is exact. The decimal 0.375 is another exact representation of the same rational number. But $1/3 = 0.333\ldots$ cannot be stored as a finite decimal. Software approximates it using a finite number of binary or decimal digits.

To add fractions, express them with a common denominator:

$$\frac{5}{12} + \frac{7}{18} = \frac{15}{36} + \frac{14}{36} = \frac{29}{36}. \tag{4.1}$$

The least common multiple of 12 and 18, namely 36, gives the smallest useful common denominator.

WORKED EXAMPLE: When sensor logs line up

A temperature sensor posts a reading every 45 s and a humidity sensor every 60 s; both posted at midnight. Factor the intervals to find the next simultaneous record:

$$45 = 3^2 \times 5, \qquad 60 = 2^2 \times 3 \times 5,$$
$$\mathrm{LCM} = 2^2 \times 3^2 \times 5 = 180\,\mathrm{s} = 3\,\mathrm{min}.$$

Check: $180/45 = 4$ and $180/60 = 3$ are whole numbers, and each hour contains $3600/180 = 20$ such paired records.

4.3 Averages answer different questions

The arithmetic mean is

$$\bar{x} = \frac{1}{n}\sum_{i=1}^{n} x_i. \tag{4.2}$$

It is appropriate when observations carry equal weight. If groups have different sizes, use a weighted mean. The median answers a different question: it is the middle ordered value and is less sensitive to extreme observations.

WORKED EXAMPLE: Average production rate

A line produces 120 parts in 2 hours and 330 parts in 3 hours. Averaging the rates directly gives $(60+110)/2 = 85$ parts/hour, but the periods are unequal. The total-rate calculation is

$$\bar{r} = \frac{120 + 330}{2 + 3} = 90/\text{h}.$$

The incorrect method weights the two-hour and three-hour periods equally. The correct method weights each rate by its duration.

PUSH THE IDEA

Computers use finite representations. In binary floating-point arithmetic, even 0.1 is generally not exact. Repeated addition can therefore accumulate tiny errors. Financial systems often store integer cents or use decimal arithmetic; measurement systems track tolerances larger than representation noise. The representation should fit the consequence of error.

COMPANION LAB: Fraction and Average Workbench

Enter fractions, decimals, and weighted groups. The lab displays exact rational results beside decimal approximations, constructs a common denominator, and compares mean, weighted mean, and median under an adjustable outlier.

CHECK YOURSELF

1. Evaluate $7/15 + 11/20$ exactly.
2. Find the greatest common factor and least common multiple of 84 and 126.
3. Find the weighted mean of 40 units at \$12 and 15 units at \$18.
4. Give a situation in which the median communicates more than the mean.

CHAPTER 5

Ratio, Scale, and Percentage

Lab 05: Scale Without Losing the Units

Original scope: Chapter V, printed pages 59–75

Powers and roots, efficient arithmetic, ratio, proportion, means, percentage, and practical scaling.

THE 1899 IDEA

Castle insists that quantities must be expressed in a common unit before they are compared. The ratio of 10 ft to 4 in is therefore $120 : 4 = 30 : 1$, not $10 : 4$. He defines a proportion as equality between ratios and uses percentage to compare changes on a common base of one hundred.

Source: Castle (1899), pp. 59–75.

5.1 Ratios compare; units measure

A ratio between like quantities is dimensionless:

$$\frac{12\,\mathrm{m}}{3\,\mathrm{m}} = 4. \tag{5.1}$$

A proportion states that two ratios are equal:

$$\frac{a}{b} = \frac{c}{d} \quad\Longrightarrow\quad ad = bc. \tag{5.2}$$

Cross-multiplication is not magic; it is ordinary multiplication of both sides by bd. Preserve the order of comparison. If one side is model divided by real, the other must be model divided by real.

5.2 Scale drawings

A scale of $1 : 50$ means one unit on the drawing represents 50 of the same units in reality:

$$L_{\text{real}} = L_{\text{drawing}} \times 50. \tag{5.3}$$

WORKED EXAMPLE: From drawing to workshop

A bracket span measures 12.5 cm on a 1 : 50 drawing.

$$\begin{aligned} L_{\text{real}} &= 12.5\,\text{cm}(50) \\ &= 625\,\text{cm} = 6.25\,\text{m}. \end{aligned}$$

Direction check: the real object must be larger than the drawing, so multiplication by 50 is plausible.

5.3 Percentage separates amount from rate

Percentage change is relative to the original value:

$$p = 100\frac{\text{new} - \text{old}}{\text{old}}. \tag{5.4}$$

Two factories may each add 500 units of output, but if one began at 2,000 and the other at 10,000, their rates of increase are 25% and 5%. Amount answers "how much?" Percentage answers "how large relative to the base?"

Successive percentage changes multiply rather than cancel. An increase of 10% followed by a decrease of 10% gives $1.10(0.90) = 0.99$, a net decrease of 1%.

A spec sheet reads the same way. A phone battery rated 4500 mA h showing 88% health holds $4500(0.88) = 3960$ mA h, and a "40% off" sale on a price first raised 25% leaves $1.25(0.60) = 0.75$ of the original: a 25% discount, not 40%. Battery gauges, load monitors, and discount banners are all percentage instruments, and the base of each percentage decides what the number means.

WORKED EXAMPLE: Range from a battery drop

An EV covers 54 km while its charge display falls from 82% to 64%. The trip consumed $82 - 64 = 18$ points of the pack, and range stands to points on both sides of the proportion:

$$\begin{aligned} \frac{R}{100} &= \frac{54}{18}, \\ R &= \frac{54}{18}(100) = 300\,\text{km}. \end{aligned}$$

Check: 18% of 300 km returns the observed 54 km; the projection assumes the remaining charge is spent at the same rate.

IN TODAY'S NOTATION

Castle uses "direct proportion" for a mean-proportional relationship in one passage. Current usage usually says y is directly proportional to x when $y = kx$. This edition uses the current convention and labels historical wording when necessary.

PUSH THE IDEA

Scale also magnifies uncertainty. If the drawing measurement is 12.5 cm plus or minus 0.1 cm, then at 1 : 50 the full-size uncertainty is plus or minus 5 cm. The result is best reported as 6.25 m plus or minus 0.05 m.

COMPANION LAB: Scale Without Losing the Units

Enter drawing length, scale denominator, and measurement uncertainty. The lab reports real length in metric and customary units, preserves the unit chain, and propagates uncertainty. Reverse mode calculates a drawing length from a real object.

CHECK YOURSELF

1. A line is 16 cm on a 1 : 25 plan. Find the real length.
2. A 12 m wall is represented at 1 : 80. Find drawing length in centimetres.
3. An output rises from 840 to 966. Find percentage increase.
4. Find the net percentage change after a 20% increase and a 20% decrease.

CHAPTER 6

Algebra as a Machine

Lab 06: Formula Rearranger

Original scope: Chapter VI, printed pages 76–95

Symbols, operations, factors, indices, simplification, substitution, and rearranging formulas for unknown quantities.

THE 1899 IDEA

Castle introduces algebra as generalized arithmetic. A letter stands for a number whose value may be unknown or allowed to vary. The same laws that govern numbers govern symbolic expressions, provided each operation is applied to the whole expression consistently.

Source: Castle (1899), pp. 76–95.

6.1 An equation preserves balance

An equation states that two expressions have the same value. Solving means rewriting that equality so the desired unknown stands alone. Any reversible operation applied to one side must be applied to the other.

For the linear equation $3x + 7 = 28$:

$$\begin{aligned} 3x + 7 &= 28, \\ 3x &= 21, \\ x &= 7. \end{aligned}$$

Substitution checks the result: $3(7) + 7 = 28$.

HISTORY THREAD: Al-Khwarizmi and algebra in words

Around the early ninth century, Muhammad ibn Musa al-Khwarizmi worked in the scholarly world centered on Baghdad. His algebra treated linear and quadratic problems connected with inheritance, trade, land measurement, and canals. It was written rhetorically: the operations were explained in words rather than our compact symbols. The title's word *al-jabr* became "algebra."

Calling this simply "Arabic mathematics" can hide the network. Arabic was a major language of scholarship, but the contributors were ethnically and religiously diverse, and they studied Greek, Indian, Persian, and earlier work. Translation was not passive storage. It was a technology for combining, criticizing, and extending ideas.

Historical note: MacTutor, "Al-Khwarizmi" and "Arabic mathematics".

6.2 Structure makes simplification possible

Like terms share the same variable factors and exponents. Thus $4x + 7x = 11x$, but $4x + 7x^2$ cannot be combined. The distributive law connects expansion and factoring:

$$a(b + c) = ab + ac, \qquad ab + ac = a(b + c). \tag{6.1}$$

Factoring exposes zeros, common structure, and opportunities to cancel. But cancellation applies to factors, not terms:

$$\frac{x^2 - x}{x} = \frac{x(x-1)}{x} = x - 1 \quad (x \neq 0). \tag{6.2}$$

The condition $x \neq 0$ is part of the result because the original denominator cannot be zero.

6.3 Indices compress repeated multiplication

For nonzero a,

$$a^m a^n = a^{m+n}, \qquad \frac{a^m}{a^n} = a^{m-n}, \qquad (a^m)^n = a^{mn}, \qquad a^{-n} = \frac{1}{a^n}. \tag{6.3}$$

Fractional powers represent roots: $a^{1/2} = \sqrt{a}$. These rules are structural, not merely typographical. They make scale, scientific notation, and logarithms manageable.

WORKED EXAMPLE: Rearrange a power formula

Electrical power is $P = V^2/R$. Solve for voltage and evaluate for $P = 48$ W and $R = 12\,\Omega$.

$$PR = V^2,$$
$$V = \sqrt{PR} = \sqrt{48 \times 12} = 24\,\text{V}.$$

The negative square root is mathematically possible for signed voltage, but the requested magnitude is positive. Context selects the branch.

WORKED EXAMPLE: Spindle speed from cutting speed

Handbook cutting speed for aluminium with a carbide bit is about $v = 75\,\text{m/min}$, and rim speed obeys $v = \pi d N$. Solve for the spindle speed N, then evaluate for a 10 mm bit, $d = 0.010\,\text{m}$:

$$N = \frac{v}{\pi d} = \frac{75}{\pi(0.010)},$$
$$N = 2387\,\text{rev/min} \approx 2400\,\text{rev/min}.$$

Substitution check: $\pi(0.010)(2387) = 75.0\,\text{m/min}$, so set the nearest speed the drill press offers.

PUSH THE IDEA

Symbolic algebra systems can rearrange and simplify expressions, but they do not know which physical branch, unit, or operating constraint matters unless it is encoded. Computer algebra accelerates transformations; the human still owns the model, assumptions, and interpretation.

COMPANION LAB: Formula Rearranger

Choose a common engineering formula, select the unknown, and watch each balance-preserving step. Substitute values with units, compare symbolic and numeric answers, and test the result in the original equation.

CHECK YOURSELF

1. Solve $5x - 13 = 42$.
2. Factor $6x^2 - 15x$ completely.
3. Simplify $(a^3b^{-2})^2/(ab^{-1})$.
4. Rearrange $s = ut + \frac{1}{2}at^2$ for a.

CHAPTER 7

Approximation with Discipline

Lab 07: Error Budget

Original scope: Chapter VII, printed pages 96–109

Approximate values, common factors and multiples, algebraic fractions, roots, significant figures, and bounded error.

THE 1899 IDEA

Castle treats approximation as an ordinary condition of practical work. Exact symbolic forms are useful, but measured quantities and hand calculations often need a controlled number of figures. The aim is not to avoid approximation; it is to know where it entered and what it can change.

Source: Castle (1899), pp. 96–109.

7.1 An approximation should carry a bound

If a length is rounded to the nearest millimetre and reported as 84 mm, the implied interval is usually

$$83.5\,\text{mm} \leq L < 84.5\,\text{mm}. \tag{7.1}$$

Absolute error has the same unit as the measurement. Relative error divides the absolute error by a reference value; percentage error multiplies relative error by 100.

7.2 Significant figures reflect scale

The measurements 0.004,20 m and 4.20 mm both have three significant figures. Leading zeros locate the decimal point; trailing zeros after a decimal may communicate precision. Scientific notation makes this explicit:

$$0.004{,}20\,\text{m} = 4.20 \times 10^{-3}\,\text{m}. \tag{7.2}$$

For addition and subtraction, decimal-place resolution controls the result. For multiplication and division, relative precision is the better guide. Keep guard digits during intermediate work and round once at the end.

WORKED EXAMPLE: Worst-case area from measured sides

A plate is $L = 250.0\,\text{mm} \pm 0.5\,\text{mm}$ by $W = 120.0\,\text{mm} \pm 0.5\,\text{mm}$. The nominal area is

$$A = LW = 30{,}000\,\text{mm}^2.$$

The first-order worst-case relative uncertainty is approximately

$$\frac{\Delta A}{A} \approx \frac{0.5}{250.0} + \frac{0.5}{120.0} = 0.00617.$$

Thus $\Delta A \approx 185\,\text{mm}^2$, so a sensible report is $A = 30{,}000\,\text{mm}^2 \pm 190\,\text{mm}^2$.

Current hardware reads the same way. A desktop 3D printer positioning to ±0.1 mm per axis cannot honor a 0.05 mm clearance in a printed hinge, whatever the slicer preview suggests, and a phone GPS fix quoted at ±3 m cannot settle a two-metre property-line question. Checking the uncertainty before trusting the number is the difference between a measurement and a guess.

WORKED EXAMPLE: Smart-plug energy budget

A smart plug meters a space heater at $P = 1.2\,\text{kW} \pm 0.1\,\text{kW}$ running $t = 3.0\,\text{h} \pm 0.1\,\text{h}$ each evening. Nominal energy is $E = Pt = 3.6\,\text{kW h}$; for a product, relative errors add:

$$\frac{\Delta E}{E} \approx \frac{0.1}{1.2} + \frac{0.1}{3.0} = 0.0833 + 0.0333 = 0.117,$$
$$\Delta E \approx 0.117(3.6) = 0.42\,\text{kW h}.$$

The wattage reading contributes 8.3 of the 11.7 percentage points, so a better plug, not a better timer, is what tightens this estimate.

7.3 Exact structure before numerical approximation

Greatest common factors, least common multiples, factoring, and cancellation should be used before decimal approximation when possible. Exact simplification reduces work and prevents avoidable rounding error. For example,

$$\frac{84}{126} = \frac{2}{3} \tag{7.3}$$

communicates structure more clearly than 0.6666667.

HISTORY THREAD: Ramanujan: intuition before proof

Srinivasa Ramanujan (1887–1920), raised in Tamil Nadu, developed astonishing results in number theory, continued fractions, and infinite series with little access to the formal research world of his time. His notebooks often recorded the destination without the route expected by professional mathematicians. Collaboration later supplied proofs, corrections, and new questions, but the original leaps remain remarkable.

Ramanujan makes the difference between discovery and verification impossible to ignore. A pattern may be seen before it can be justified. Proof does not replace intuition; it tests, communicates, and stabilizes it. Practical work has the same rhythm: estimate first, calculate carefully, then ask whether the answer survived contact with evidence.

Historical note: MacTutor, "Srinivasa Ramanujan".

PUSH THE IDEA

An error budget lists sources of uncertainty and their influence on the final quantity. Sensitivity is measured by how much the output changes for a small input change. A model may be numerically precise yet practically useless if it is highly sensitive to an input that cannot be measured reliably.

COMPANION LAB: Error Budget

Enter a formula, nominal inputs, and input uncertainties. The lab compares worst-case and root-sum-square estimates, highlights the dominant contributor, and shows how premature rounding changes the reported result.

CHECK YOURSELF

1. State the interval implied by 18.2 cm rounded to the nearest millimetre.
2. Find the percentage error when 49.8 Ω is used for a true 50.0 Ω resistance.
3. Round 0.0098765 to three significant figures.
4. Estimate worst-case relative uncertainty in $Q = ab/c$ from independent bounds on a, b, c.

PART II

Symbols, Graphs, and Representation

CHAPTER 8

Unknowns in Balance

Lab 08: Equation Balance

Original scope: Chapter VIII, printed pages 110–127

Linear equations, simultaneous equations, elimination, substitution, intersections, and model constraints.

THE 1899 IDEA

Castle moves from symbolic manipulation to equations that describe practical conditions. A single equation determines one unknown; two independent equations can determine two unknowns. He emphasizes checking by substitution, an essential habit whether the solution is found by hand or by software.

Source: Castle (1899), pp. 110–127.

8.1 An unknown becomes knowable through constraints

The equation $2x + 5 = 19$ contains one unknown and one independent constraint. Two unknowns generally need two independent equations. Consider

$$2x + y = 11, \tag{8.1}$$
$$x - y = 1. \tag{8.2}$$

Adding the equations eliminates y, giving $3x = 12$, so $x = 4$ and $y = 3$. Substitution into both original equations confirms the pair.

HISTORY THREAD: Brahmagupta gives zero rules

In seventh-century India, Brahmagupta tried to state a consistent arithmetic for zero and negative quantities. He described positive numbers as fortunes and negative numbers as debts, defined zero through subtracting a number from itself, and gave rules for addition and multiplication. Some of his proposed rules for division by zero were wrong, but that failure is historically valuable: extending a number system exposes questions the old rules never had to answer.

Zero is both a placeholder in notation and a number in arithmetic. Joining those roles was a conceptual achievement, not an empty symbol. Mathematics grows when an exception becomes an object that can be reasoned about.

Historical note: MacTutor, "Brahmagupta" and "Zero".

8.2 Equations are intersecting models

Each linear equation in x and y is a straight line. A unique solution is their intersection. Parallel distinct lines have no solution; coincident lines have infinitely many. The algebraic symptoms are equally important:

- a true statement such as $0 = 0$ indicates dependent equations;
- a contradiction such as $0 = 7$ indicates inconsistent equations;
- a numerical pair indicates a unique intersection.

8.3 Model the words before solving

Let the variables represent quantities with units, and state the meaning before writing equations. This prevents a correct algebraic solution to the wrong problem.

WORKED EXAMPLE: Blend two stock solutions

How many litres of 20% and 50% solutions make 30 L at 32%? Let x be litres of 20% solution and y litres of 50% solution.

$$x + y = 30,$$
$$0.20x + 0.50y = 0.32(30) = 9.6.$$

Substitute $x = 30 - y$:

$$0.20(30 - y) + 0.50y = 9.6,$$
$$0.30y = 3.6,$$
$$y = 12, \qquad x = 18.$$

Check: $0.20(18) + 0.50(12) = 9.6$ litres of solute.

The same two-line model prices modern services. A cloud host charging \$12 monthly plus \$0.02 per build and a rival charging \$30 flat break even where $12 + 0.02n = 30$, so $n = 900$ builds per month. Below the crossover the metered plan wins; above it the flat plan does. Phone data, freight, and print-on-demand quotes reduce to the same pair of intersecting lines.

WORKED EXAMPLE: Wind from two ground speeds

A survey quadcopter flies a straight leg downwind and returns along the same track upwind. GPS logs ground speeds of 16.0 m/s out and 9.5 m/s back. Let a be airspeed and w wind speed.

$$\begin{aligned} a + w &= 16.0, \\ a - w &= 9.5, \\ 2a &= 25.5, \\ a &= 12.75\,\text{m/s}, \qquad w = 3.25\,\text{m/s}. \end{aligned}$$

Check: $12.75 + 3.25 = 16.0$ and $12.75 - 3.25 = 9.5$. As w approaches a, the return leg's ground speed approaches zero; the battery, not the algebra, sets that limit.

IN TODAY'S NOTATION

Equations do not enforce physical admissibility automatically. A negative tank volume or a mixture fraction above 100% may satisfy manipulated symbols after a modeling mistake. Units, signs, and feasible ranges are additional constraints.

PUSH THE IDEA

Large systems are written as $A\mathbf{x} = \mathbf{b}$. Numerical linear algebra solves thousands or millions of coupled equations, but conditioning matters. If small input changes cause large solution changes, more digits do not repair the model. Geometry explains the warning: nearly parallel constraints intersect at a poorly determined point.

COMPANION LAB: Equation Balance

Enter two linear equations. The lab shows elimination and substitution steps, plots both lines, classifies the system, and tests the result. A sensitivity slider makes the lines nearly parallel to reveal an ill-conditioned solution.

CHECK YOURSELF

1. Solve $7x + 4 = 3x + 28$.
2. Solve $3x + 2y = 19$ and $x - y = 3$.
3. Classify $2x + 4y = 8$ and $x + 2y = 5$.
4. **Write, but do not solve, equations for 15 items consisting of $8 and $13 parts totaling $155.**

CHAPTER 9

Graphs as Measuring Instruments

Lab 09: Plot and Measure

Original scope: Chapter IX, printed pages 128–141

Coordinate graphs, scales, graphical arithmetic, slopes, interpolation, and right triangles.

THE 1899 IDEA

Castle treats a graph as a practical calculator. Lengths laid off to a chosen scale can add vectors, solve proportions, interpolate tables, and expose the geometry of right triangles. The drawing is not decoration; it is an analog representation whose accuracy depends on its scale and construction.

Source: Castle (1899), pp. 128–141.

9.1 A scale maps numbers to positions

On a graph, the horizontal and vertical axes may use different scales. Every plotted point represents an ordered pair (x, y). A straight-line relation has the form

$$y = mx + b, \tag{9.1}$$

where $m = \Delta y/\Delta x$ is slope and b is the vertical intercept. Slope has units whenever the axes have units. On a distance-time graph it is speed; on a force-extension graph it may be stiffness.

HISTORY THREAD: Descartes joins two languages

Rene Descartes's *La Geometrie* (1637) is associated with the decisive joining of algebra and geometry: a curve could be described by an equation, and an equation could be investigated as a curve. Earlier thinkers had used coordinate-like ideas, and later editors and teachers helped make Cartesian methods systematic. The durable contribution was therefore not a lone flash, but a language that a community learned to use.

A coordinate graph is a translation device. Shape becomes symbol; symbol becomes shape. Good mathematical representations do not merely display an answer. They let one kind of reasoning check another.

Historical note: MacTutor, "Rene Descartes" and "Frans van Schooten".

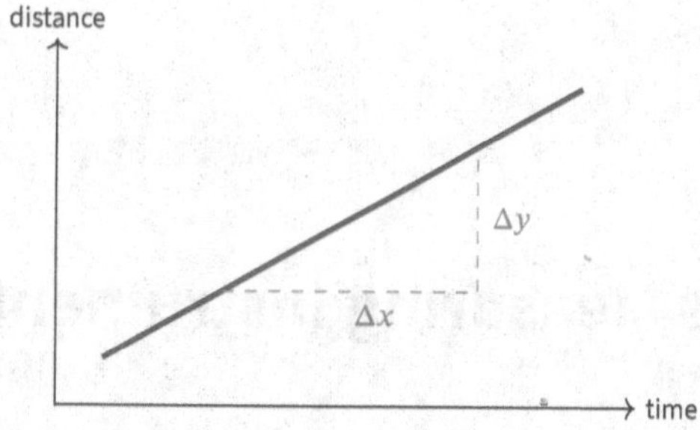

9.2 Interpolation belongs inside the data range

Linear interpolation between (x_1, y_1) and (x_2, y_2) estimates

$$y = y_1 + \frac{x - x_1}{x_2 - x_1}(y_2 - y_1). \tag{9.2}$$

Interpolation assumes the relation is adequately straight over that interval. Extrapolation extends beyond observed data and is riskier because the trend may change.

WORKED EXAMPLE: Interpolate a thermistor table

A 3D printer reads its hot-end thermistor at 4.50 kΩ. The sensor table lists 5.32 kΩ at 60 °C and 3.59 kΩ at 70 °C. Interpolate with resistance as x:

$$\begin{aligned} T &= 60 + \frac{4.50 - 5.32}{3.59 - 5.32}(70 - 60) \\ &= 60 + \frac{-0.82}{-1.73}(10) = 64.7\,^{\circ}\mathrm{C}. \end{aligned}$$

The true curve is exponential, but over one ten-degree bracket the straight chord serves. The reading sits inside the interval, where interpolation belongs; firmware repeats this lookup many times each second.

9.3 Right triangles connect components

For perpendicular components a and b, the resultant length is

$$c = \sqrt{a^2 + b^2}. \tag{9.3}$$

The same geometry describes a diagonal brace, displacement components, vector magnitude, and distance between coordinate points.

WORKED EXAMPLE: Ramp geometry

A ramp rises 0.75 m over a horizontal run of 6.0 m.

$$L = \sqrt{6.0^2 + 0.75^2} = 6.047\ \text{m},$$

$$\text{grade} = 100\frac{0.75}{6.0} = 12.5\%.$$

The Pythagorean result is the ramp length; the percentage is rise per horizontal run. Neither is the angle, which is developed in Chapter 11. Accessibility codes make the percentage a legal object: a wheelchair ramp is commonly limited to an 8.33% grade, one unit of rise per twelve of run, so this 12.5% ramp fails inspection until its run lengthens to $12(0.75) =$ 9.0 m.

PUSH THE IDEA

Plotting residuals, the differences between measurements and model predictions, often reveals more than plotting the fitted line. Curvature, cycles, and outliers in residuals expose missing structure. A modern graph is still a measuring instrument; its target is now often the model's error.

COMPANION LAB: Plot and Measure

Enter or drag points on a coordinate plane. Fit a line, read slope with units, interpolate inside the data, and extrapolate beyond it. The lab distinguishes data points, the model, and residuals, then switches to a right-triangle mode for component geometry.

CHECK YOURSELF

1. Find the slope through $(2, 5)$ and $(8, 20)$.
2. Interpolate at $x = 7$ between $(5, 18)$ and $(9, 30)$.
3. Find the diagonal of a 9 m by 12 m rectangle.
4. Explain why extrapolating a straight trend far beyond the data is risky.

CHAPTER 10

Logarithms: Multiplication by Addition

Lab 10: Logarithm Lens

Original scope: Chapter X, printed pages 142–157

Powers of ten, logarithms, characteristics and mantissas, products, quotients, roots, and scale compression.

THE 1899 IDEA

Castle presents common logarithms as a labor-saving technology. Products become sums, quotients become differences, and powers become multiplications. Tables separate a logarithm into characteristic and mantissa so long calculations can be performed with lookup, addition, and interpolation.

Source: Castle (1899), pp. 142–157.

10.1 A logarithm answers an exponent question

The statement

$$\log_b x = y \tag{10.1}$$

means exactly $b^y = x$. Common logarithms use base 10; natural logarithms use base e. The inverse relationships are

$$b^{\log_b x} = x, \qquad \log_b(b^y) = y. \tag{10.2}$$

Only positive real numbers have real logarithms. The argument of a logarithm is dimensionless; in physical formulas, a logarithm applies to a ratio such as x/x_0 even when the reference is sometimes left implicit.

10.2 The laws follow from exponent laws

Because $b^m b^n = b^{m+n}$,

$$\log_b(xy) = \log_b x + \log_b y,$$
$$\log_b(x/y) = \log_b x - \log_b y,$$
$$\log_b(x^p) = p \log_b x.$$

These identities powered Castle's tables and still power algorithms, data transformations, and analysis of exponential processes.

WORKED EXAMPLE: Compound growth by logarithms

How long does an investment growing at 6% annually take to double?

$$2 = (1.06)^n,$$
$$\ln 2 = n \ln(1.06),$$
$$n = \frac{\ln 2}{\ln 1.06} = 11.90.$$

The balance first exceeds double after 12 annual compounding periods. The result also explains the "rule of 72": $72/6 \approx 12$ years.

WORKED EXAMPLE: Size of a cryptographic keyspace

A 256-bit encryption key can take 2^{256} values. How many decimal digits is that? Apply the power law to the common logarithm:

$$\log_{10}(2^{256}) = 256 \log_{10} 2 = 256(0.30103) = 77.064,$$
$$2^{256} = 10^{0.064} \times 10^{77} \approx 1.16 \times 10^{77}.$$

The characteristic 77 means 78 digits; the mantissa supplies the leading figures. A rough check from $2^{10} \approx 10^3$ gives $10^{3\times 25.6} = 10^{76.8}$, which agrees. Brute force fails on the exponent alone.

10.3 Logarithmic scales preserve ratios

Equal distances on a linear axis represent equal differences. Equal distances on a logarithmic axis represent equal ratios. The step from 1 to 10 occupies the same distance as 10 to 100. This compression allows many orders of magnitude to appear together and converts exponential curves into straight lines.

Decibels apply the ratio principle directly: a power ratio in decibels is $10 \log_{10}(P_2/P_1)$. Driving a speaker from 85 dB to 91 dB multiplies its power by $10^{6/10} = 10^{0.6} \approx 4$, which is why the last few decibels of volume drain a battery so fast. Sound meters, audio apps, and Wi-Fi signal readouts in dBm report logarithms because the underlying ratios span orders of magnitude.

IN TODAY'S NOTATION

Log tables usually supplied four or five reliable figures, which made their precision visible. Calculators display more digits, but the input measurements still govern meaningful precision. A longer display is not a better measurement.

PUSH THE IDEA

Decibels, pH, earthquake magnitude, information measures, and algorithmic complexity use logarithmic ideas. Logarithms are especially useful when the question concerns ratios, repeated multiplication, or orders of magnitude.

COMPANION LAB: Logarithm Lens

Move a value across linear and logarithmic axes. See products become horizontal shifts, compare exponential data before and after a log transform, and replay a table-based multiplication beside the calculator result.

CHECK YOURSELF

1. Evaluate $\log_{10}(100000)$ and $\log_2(1/8)$.
2. Solve $3^x = 50$ using logarithms.
3. Expand $\ln(a^2\sqrt{b}/c)$.
4. Why are logarithmic axes inappropriate for zero or negative values in an ordinary real plot?

CHAPTER 11

Triangles in the Field

Lab 11: Indirect Measure

Original scope: Chapter XI, printed pages 158–177

Sine, cosine, tangent, inverse functions, heights and distances, bearings, and measurement sensitivity.

THE 1899 IDEA

Castle defines the trigonometric functions as ratios of sides in a right triangle and immediately applies them to heights and distances. Tables convert between an angle and its function. The practical purpose is indirect measurement: determine what cannot be reached from quantities that can.

Source: Castle (1899), pp. 158–177.

11.1 Three ratios organize a right triangle

For angle θ in a right triangle,

$$\sin\theta = \frac{\text{opposite}}{\text{hypotenuse}}, \qquad \cos\theta = \frac{\text{adjacent}}{\text{hypotenuse}}, \qquad \tan\theta = \frac{\text{opposite}}{\text{adjacent}}. \tag{11.1}$$

The inverse functions recover an angle from a ratio, for example $\theta = \tan^{-1}(h/d)$. The superscript -1 here means inverse function, not reciprocal; $\tan^{-1} x$ is not $1/\tan x$.

HISTORY THREAD: A sine table crosses languages

The modern sine grew through a chain of translation and recalculation. Indian mathematicians worked with half-chords - ancestors of the sine function - for astronomy. Scholars writing in Arabic expanded trigonometric tables and methods. In the thirteenth century, Nasir al-Din al-Tusi presented plane and spherical trigonometry as a subject in its own right, not merely a chapter of astronomy.

The chain matters as much as the names. A translated term can change meaning; a table recomputed in a new observatory can gain precision; a tool borrowed for one purpose can become an independent theory. Knowledge survives by being used, and it grows by changing hands.

Historical note: MacTutor, “Nasir al-Din al-Tusi” and “The trigonometric functions”.

11.2 Choose the ratio containing what you know

Sketch and label the triangle before selecting a formula. If the known sides are opposite and adjacent to the angle, tangent is direct. If the hypotenuse is involved, sine or cosine may fit. The mnemonic is less important than correctly identifying the sides relative to the chosen angle.

WORKED EXAMPLE: Height of an inaccessible structure

An observer stands 42.0 m from a tower. The angle of elevation to the top is 31.5°, and the instrument is 1.60 m above ground.

$$h_{\text{above instrument}} = 42.0\tan(31.5^\circ) = 25.74\,\text{m},$$
$$h_{\text{tower}} = 25.74 + 1.60 = 27.34\,\text{m}.$$

A sensible report is 27.3 m. The instrument height matters because the triangle begins at the line of sight, not at ground level.

WORKED EXAMPLE: Space solar rows for winter sun

A ground-mount solar panel 1.65 m long is tilted at 32°. At winter noon the sun stands 25° above the horizon. Each row needs its own footprint plus the shadow it throws.

$$\text{rise} = 1.65\sin(32^\circ) = 0.874\,\text{m},$$
$$\text{footprint} = 1.65\cos(32^\circ) = 1.399\,\text{m},$$
$$\text{shadow} = \frac{0.874}{\tan(25^\circ)} = 1.875\,\text{m}.$$

Spacing rows $1.399 + 1.875 = 3.27$ m apart keeps every panel lit at the year's lowest useful sun. One layout uses all three ratios.

11.3 Angles amplify error unevenly

For $h = d \tan \theta$, sensitivity to angle is

$$\frac{\partial h}{\partial \theta} = d \sec^2 \theta, \tag{11.2}$$

with θ measured in radians. As θ approaches 90°, the sensitivity grows rapidly. A small angular error can then produce a large height error. A good survey geometry avoids extremely shallow or steep sight angles when possible.

PUSH THE IDEA

Triangulation generalizes from one right triangle to networks of angles and distances. GPS, photogrammetry, robotic localization, and computer vision all estimate position from geometric constraints, then combine redundant observations to reduce uncertainty.

COMPANION LAB: Indirect Measure

Adjust baseline distance, viewing angle, instrument height, and angular uncertainty. The lab draws the triangle, calculates height, and displays a sensitivity band. Compare geometries to find a safer field setup.

CHECK YOURSELF

1. A cable of length 18 m rises at 24°. Find vertical rise.
2. Find the angle of a ramp rising 0.8 m over 9.5 m horizontally.
3. From 60 m away, a top is sighted at 18° from an instrument 1.5 m high. Find total height.
4. Explain why angle units must match calculator mode.

CHAPTER 12

The Slide Rule and the Shape of Computation

Lab 12: Virtual Slide Rule

Original scope: Chapter XII, printed pages 178–188

Logarithmic scales, significant figures, range estimation, analog computation, and the design of calculational tools.

THE 1899 IDEA

Castle's slide rule chapter turns logarithm laws into a physical instrument. Distances proportional to logarithms allow multiplication by adding lengths. The rule supplies significant digits; the operator must place the decimal point by estimation. The tool therefore combines mechanism with judgment.

Source: Castle (1899), pp. 178–188.

12.1 A logarithmic ruler adds exponents

On a log scale, the coordinate of x is proportional to $\log x$. Placing the start of a sliding scale at a and reading beneath b produces a total distance proportional to

$$\log a + \log b = \log(ab). \tag{12.1}$$

Division reverses the displacement. Square and cube scales encode multiplication of logarithms by 2 and 3. The slide rule is a physical user interface for algebraic identities.

12.2 The missing decimal point is a feature

A typical rule may show that 2.4×3.1 begins with 7.44, but not whether the answer is 0.744, 7.44, or 74.4. Estimation supplies order of magnitude. This forced the operator to maintain a mental model of scale, making gross errors easier to catch.

WORKED EXAMPLE: Multiply with mantissas and estimate the range

Estimate 238×0.047. Separate powers of ten:

$$238 \times 0.047 = (2.38 \times 10^2)(4.7 \times 10^{-2}) = (2.38 \times 4.7) \times 10^0.$$

The slide rule gives approximately 11.2. Mental bounds confirm it: $200 \times 0.05 = 10$. A calculator gives 11.186, but the input 0.047 supports only about two significant figures, so 11 may be the honest report.

WORKED EXAMPLE: Audit a confident answer

An AI assistant, asked what energy a phone battery rated 3.7 V and 4.5 A h stores, replies 16.7 kW h. Separate significand from exponent before believing it:

$$E = 3.7 \times 4.5 = 16.65 \approx 17\,\text{W h}, \qquad \text{mental bound: } 4 \times 4 = 16.$$

The digits are right and the exponent is off by 10^3: 16.7 kW h is an electric-car pack, not a phone. The machine supplied confident figures; placing the decimal point is still the operator's job.

12.3 Tools shape habits

Different tools expose different structure. Written arithmetic exposes carries and place value. A slide rule exposes ratios and significant figures. A spreadsheet exposes dependencies between cells. Code exposes named operations, branching, and repeatability. A calculator exposes results quickly but may hide the chain that produced them.

IN TODAY'S NOTATION

Use the least powerful tool that leaves the reasoning visible enough to check. The goal is not nostalgia for slow calculation. It is preserving scale, assumptions, and auditability while gaining speed.

PUSH THE IDEA

Floating-point numbers also store a significand and an exponent. Their range is vast, but their precision is finite. Overflow, underflow, and cancellation are digital descendants of the range and significant-figure concerns a slide-rule operator handled consciously.

COMPANION LAB: Virtual Slide Rule

Drag C and D scales to multiply and divide. The display hides the decimal point until you supply an order-of-magnitude estimate. A second mode shows the same operation as logarithm addition and as floating-point significand plus exponent.

CHECK YOURSELF

1. **Estimate 6.2×0.083 before calculating.**
2. **Use powers of ten to explain the decimal location in 4700/0.023.**
3. **Why does a slide rule usually provide fewer trustworthy digits than a digital calculator displays?**
4. **Name one advantage and one risk of hiding intermediate steps in software.**

CHAPTER 13

Sequences, Growth, and Finance

Lab 13: Growth Engine

Original scope: Chapter XIII, printed pages 189–199

Arithmetic and geometric progressions, means, simple and compound interest, discount, and exponential growth.

THE 1899 IDEA

Castle connects sequences to practical finance. Equal additive steps form an arithmetic progression; equal multiplicative steps form a geometric progression. Simple interest adds a fixed amount, while compound interest multiplies the balance repeatedly. The difference becomes dramatic over time.

Source: Castle (1899), pp. 189–199.

13.1 Additive and multiplicative change

An arithmetic sequence has constant difference d:

$$a_n = a_1 + (n-1)d. \tag{13.1}$$

A geometric sequence has constant ratio r:

$$a_n = a_1 r^{n-1}. \tag{13.2}$$

Linear growth adds the same amount each period. Exponential growth adds the same percentage of a changing base. Confusing the two is a common source of bad forecasts.

13.2 Finite sums

The sum of n arithmetic terms is

$$S_n = \frac{n}{2}\left(2a_1 + (n-1)d\right). \tag{13.3}$$

For a geometric sequence with $r \neq 1$,

$$S_n = a_1 \frac{1-r^n}{1-r}. \tag{13.4}$$

These formulas turn repeated entries into one auditable calculation.

WORKED EXAMPLE: Compare simple and compound interest

Invest \$8,000 for 6 years at 5% annually.
Simple interest:

$$A = 8000(1 + 0.05 \times 6) = \$10{,}400.$$

Annual compounding:

$$A = 8000(1.05)^6 = \$10{,}720.77.$$

The \$320.77 difference is interest earned on earlier interest.

WORKED EXAMPLE: Sixty automatic deposits

An auto-invest app moves \$200 into a savings plan at each month's end, and the plan earns 0.4% monthly. After 5 years the accumulated deposits form a geometric sum with $a_1 = 200$, $r = 1.004$, and $n = 60$:

$$S_{60} = 200\,\frac{1.004^{60} - 1}{1.004 - 1} = 200\,\frac{0.27064}{0.004} = \$13{,}532.04.$$

Deposits alone total \$12,000; the further \$1,532.04 accrues because the first deposit compounds 59 times and the last not at all. The app's balance projection runs exactly this sum.

13.3 Present value and discount

If a future amount F grows at rate i for n periods, its present value is

$$P = \frac{F}{(1+i)^n}. \tag{13.5}$$

The calculation runs compound growth backward. The rate and period must match: a monthly rate requires a number of months, while an annual rate requires years unless a conversion convention is stated.

Compounding frequency is where modern credit quietly bites. A card quoting 24% APR compounds monthly at 2%, so a carried balance grows by $(1.02)^{12} - 1 = 26.8\%$ in a year, not 24%. The advertised figure is the nominal rate; the effective annual rate is what the balance actually does. Buy-now-pay-later offers, savings yields, and loan disclosures all turn on this same distinction.

PUSH THE IDEA

Real financial models add taxes, fees, inflation, irregular cash flows, and uncertain returns. A scenario model should separate assumptions from formulas. Monte Carlo simulation can show a distribution of outcomes, but it cannot make an unjustified return assumption trustworthy.

COMPANION LAB: Growth Engine

Compare arithmetic and geometric sequences, simple and compound interest, and present value. Change compounding frequency and inflation; inspect both a table and a graph. A cash-flow mode computes net present value from dated amounts.

CHECK YOURSELF

1. Find the 20th term of 7, 11, 15, ...
2. Find the sum of the first 12 terms of 3, 6, 12, ...
3. Find the future value of $5,000 at 4% annually for 8 years.
4. Find the present value of $10,000 due in 5 years at a 6% discount rate.

CHAPTER 14

Projection: Describing Three Dimensions on Two

Lab 14: Projection Room

Original scope: Chapter XIV, printed pages 200–227

Orthographic views, coordinate planes, true length, inclination, hidden geometry, and transformations.

THE 1899 IDEA

Castle develops orthographic projection as a disciplined method for representing points, lines, and planes on mutually perpendicular drawing planes. Multiple views recover information lost by any single projection. Rotation and auxiliary views reveal true lengths and inclinations.

Source: Castle (1899), pp. 200–227.

14.1 Projection deliberately discards one dimension

An orthographic projection maps a three-dimensional point (x, y, z) to a view. The top view may retain (x, y) and discard z; the front view may retain (x, z) and discard y. Neither view is complete, but together they constrain the point.

For endpoints $P_1 = (x_1, y_1, z_1)$ and $P_2 = (x_2, y_2, z_2)$, true length is

$$L = \sqrt{(x_2 - x_1)^2 + (y_2 - y_1)^2 + (z_2 - z_1)^2}. \tag{14.1}$$

A projected length cannot exceed the true length. Equality occurs when the line is parallel to the projection plane.

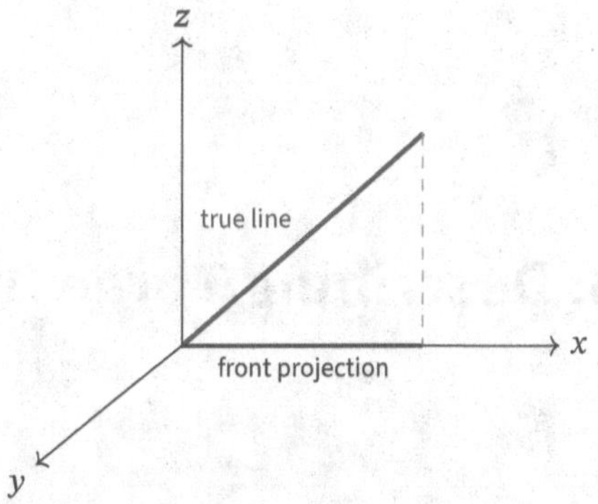

14.2 Views must remain aligned

Construction lines transfer coordinates between front, top, and side views. Alignment is not cosmetic; it is the mechanism that shows the views describe one object. Hidden lines communicate edges obscured in a view, while line type and weight create a hierarchy of evidence.

WORKED EXAMPLE: True length of a diagonal brace

A brace changes 1.8 m in x, 0.9 m in y, and 1.2 m in z.

$$L = \sqrt{1.8^2 + 0.9^2 + 1.2^2} = \sqrt{5.49} = 2.343\,\text{m}.$$

Its plan-view length is $\sqrt{1.8^2 + 0.9^2} = 2.012\,\text{m}$, shorter than the true length as expected.

14.3 Projection is a matrix operation

In computer graphics, points are multiplied by transformation matrices. Orthographic projection deletes depth after rotation into camera coordinates; perspective projection also divides by depth, making distant objects appear smaller. Castle's drawing operations are the geometric ancestors of a rendering pipeline.

WORKED EXAMPLE: True size from a drone roof survey

A drone's straight-down photo is a plan view: it keeps (x, y) and discards z, so sloped surfaces are foreshortened. The image of a roof pitched at 32° shows a horizontal rafter run of 4.6 m and a ridge of 9.2 m.

$$s = \frac{4.6}{\cos 32^\circ} = 5.424\,\text{m},$$
$$A_{\text{true}} = 9.2(5.424) = 49.90\,\text{m}^2 \quad \text{against} \quad 9.2(4.6) = 42.32\,\text{m}^2 \text{ in plan.}$$

The ridge lies parallel to the image plane and projects at true length; only the up-slope dimension shrinks. Measuring the roof from its plan area would undersell it by about 15%.

PUSH THE IDEA

CAD adds constraints, parametric dimensions, layers, and model history. Yet a model can remain ambiguous or overconstrained. The old discipline still helps: identify which view carries each dimension, keep correspondences aligned, and verify true size rather than measuring an arbitrary projection.

COMPANION LAB: Projection Room

Rotate a 3D line or block and watch top, front, and side views update. Toggle hidden lines, compare projected and true lengths, and construct an auxiliary view perpendicular to a selected line.

CHECK YOURSELF

1. Find the true distance between $(1, 2, 3)$ and $(5, 5, 9)$.
2. Which coordinate is discarded in a top view using the xy plane?
3. State when a projected line has its true length.
4. Explain one difference between orthographic and perspective projection.

PART III

Shape, Motion, and Structure

CHAPTER 15

Area by Structure

Lab 15: Shape Builder

Original scope: Chapter XV, printed pages 228–250

Plane figures, triangles, quadrilaterals, circles, sectors, annuli, ellipses, decomposition, and dimensional checks.

THE 1899 IDEA

Castle assembles the formulas of plane mensuration around constructions and decompositions. Rectangles lead to parallelograms and triangles; circles lead to sectors and annuli. The formulas are not isolated facts but consequences of rearranging shapes whose area is already known.

Source: Castle (1899), pp. 228–250.

15.1 Area has a dimensional signature

Every valid area formula has dimensions of length squared. For a triangle,

$$A = \frac{1}{2}bh. \tag{15.1}$$

Both b and h must be perpendicular lengths. The product bh has square units, and the factor 1/2 is dimensionless. Dimensional analysis cannot prove the coefficient is correct, but it can reject $A = b + h$ immediately.

HISTORY THREAD: Before and beyond Pythagoras

The rule $a^2 + b^2 = c^2$ carries Pythagoras's name, yet surviving Old Babylonian tablets show sophisticated right-triangle calculations more than a thousand years before he lived. Related results and proofs also appear in Indian and Chinese mathematical traditions. The Greek tradition made proof itself a central form of mathematical knowledge, but it should not erase the earlier computational knowledge or later independent arguments.
This distinction is philosophical as well as historical. Knowing that a rule works across many examples is not the same as proving why it must always work. Both are valuable: computation discovers and tests; proof explains necessity. The theorem's real biography belongs to many cultures.

Historical note: MacTutor, "Babylonian Pythagoras" and "Pythagoras".

15.2 Build unfamiliar shapes from familiar ones

A composite area can be found by addition and subtraction:

$$A_{\text{total}} = \sum A_{\text{added}} - \sum A_{\text{removed}}. \tag{15.2}$$

Choose a decomposition that uses measured dimensions directly and avoids subtracting nearly equal large areas when possible. The latter magnifies relative error in the small remainder.

For a circle, sector, annulus, and ellipse:

$$\begin{aligned} A_{\text{circle}} &= \pi r^2, \\ A_{\text{sector}} &= \frac{1}{2} r^2 \theta \quad (\theta \text{ in radians}), \\ A_{\text{annulus}} &= \pi(R^2 - r^2), \\ A_{\text{ellipse}} &= \pi ab, \end{aligned}$$

where a and b are ellipse semiaxes.

WORKED EXAMPLE: Plate with two holes

A rectangular plate is 420 mm by 260 mm and contains two holes of diameter 48 mm.

$$\begin{aligned} A_{\text{rectangle}} &= 420(260) = 109{,}200\ \text{mm}^2, \\ A_{\text{holes}} &= 2\pi(24)^2 = 3619\ \text{mm}^2, \\ A_{\text{net}} &= 109200 - 3619 = 105{,}581\ \text{mm}^2. \end{aligned}$$

Rounded to the measurement precision, the net area is about 105,600 mm^2, or 0.1056 m^2.

WORKED EXAMPLE: Paint for a gable-end wall

A gable-end wall is 6.8 m wide and 2.4 m to the eaves, with a gable rising a further 1.7 m. A door 0.9 m by 2.0 m and two windows 1.2 m by 1.0 m are not painted.

$$\begin{aligned} A_{\text{rectangle}} &= 6.8(2.4) = 16.32\,\text{m}^2, \\ A_{\text{gable}} &= \frac{1}{2}(6.8)(1.7) = 5.78\,\text{m}^2, \\ A_{\text{openings}} &= 0.9(2.0) + 2(1.2)(1.0) = 4.20\,\text{m}^2, \\ A_{\text{net}} &= 16.32 + 5.78 - 4.20 = 17.90\,\text{m}^2. \end{aligned}$$

Two coats mean 35.8 m^2 of coverage; at 11 m^2/L that is 3.3 L, so a single 4 L can does both coats.

15.3 Perimeter is not area

Scaling every length by k multiplies perimeter by k and area by k^2. This square-cube pattern explains why similar designs do not preserve material, heat loss, capacity, or weight in simple proportion to a linear dimension.

PUSH THE IDEA

Computational geometry represents a boundary as ordered vertices. The shoelace formula calculates polygon area directly from coordinates, while constructive solid geometry combines primitive shapes by union, intersection, and difference. These are programmable versions of Castle's decomposition method.

COMPANION LAB: Shape Builder

Assemble rectangles, triangles, circles, sectors, and holes on a plane. The lab tracks net area, perimeter, centroid estimate, unit conversion, and cost per area. A scale control demonstrates linear versus squared growth.

CHECK YOURSELF

1. Find the area of a triangle with base 14 m and perpendicular height 8.5 m.
2. Find the area of an annulus with outer radius 90 mm and inner radius 65 mm.
3. A similar figure doubles every length. By what factors do perimeter and area change?
4. Explain why subtracting two nearly equal large areas can give an uncertain small result.

CHAPTER 16

Area from Samples

Lab 16: Simpson's Rule Studio

Original scope: Chapter XVI, printed pages 251–263

Trapezoidal and Simpson rules, planimeters, numerical integration, convergence, and error estimation.

THE 1899 IDEA

Castle turns irregular boundaries into numerical area by measuring ordinates at regular intervals. The trapezoidal rule joins adjacent samples with straight lines; Simpson's rule fits parabolic arcs across pairs of intervals. A planimeter performs a related integration mechanically.

Source: Castle (1899), pp. 251-263.

16.1 Integration is accumulated contribution

For sampled values $y_i = f(x_i)$ at uniform spacing h, the composite trapezoidal rule is

$$T_n = h\left(\frac{y_0 + y_n}{2} + \sum_{i=1}^{n-1} y_i\right). \tag{16.1}$$

Composite Simpson's rule requires an even number n of intervals:

$$S_n = \frac{h}{3}\left[y_0 + y_n + 4\sum_{i\ \text{odd}} y_i + 2\sum_{i\ \text{even},\, i\neq 0,n} y_i\right]. \tag{16.2}$$

The alternating weights encode a sequence of quadratic interpolants.

HISTORY THREAD: Approaching the exact by repeated steps

In third-century China, Liu Hui refined inscribed polygons to estimate the area of a circle and gave reasoning for the limiting process. More than a thousand years later in Kerala, southwestern India, Madhava developed infinite series for trigonometric quantities and ways to correct their remainders. Their methods differ, but both turn a philosophical puzzle into a working method: how can finitely performed calculations approach an exact value defined by an infinite process?

Numerical integration inherits that tension. We never take infinitely many samples in the workshop or computer. We refine, compare, bound the remaining error, and decide when the approximation is fit for purpose.

Historical note: MacTutor, "Liu Hui" and "Madhava".

WORKED EXAMPLE: Integrate a measured profile

Suppose a profile has ordinates $[0, 1, 4, 9, 16]$ at unit spacing, corresponding to $y = x^2$ on $0 \leq x \leq 4$.

Trapezoidal rule:

$$T_4 = 1\left[\frac{0+16}{2} + 1 + 4 + 9\right] = 22.$$

Simpson's rule:

$$S_4 = \frac{1}{3}[0 + 16 + 4(1+9) + 2(4)] = \frac{64}{3}.$$

The exact integral is $\int_0^4 x^2\,dx = 64/3$, so Simpson's rule is exact here because it integrates quadratics exactly.

16.2 Refine and compare

An approximate value without an error investigation is incomplete. Repeat the calculation with half the spacing and compare. If the values stabilize, the difference provides evidence of convergence. Sharp corners, discontinuities, or sparse measurements may require local refinement rather than uniform sampling everywhere.

IN TODAY'S NOTATION

The formula does not know whether the samples are trustworthy. Measurement noise, wrong ordering, nonuniform spacing, and omitted geometry can dominate the formal integration error.

16.3 Power integrates to energy

The sampling rules are working calculus. If $E(t)$ is the energy a battery has absorbed by time t, instantaneous power is the derivative $P = dE/dt$, and accumulated energy is the integral

$$E = \int_{t_0}^{t_1} P(t)\,dt. \tag{16.3}$$

The fundamental theorem of calculus states that the two operations undo one another: differentiate the energy record and the power curve returns. When $P(t)$ arrives as telemetry samples rather than as a formula, the trapezoidal and Simpson rules perform the integral numerically.

WORKED EXAMPLE: Energy from a fast-charge session

A direct-current fast charger logs $P = [24, 96, 144, 132, 108, 84, 60]$ kW at five-minute intervals while an electric vehicle charges for 30 min: power ramps up, peaks, then tapers as the battery fills. With $h = 5\text{ min} = \frac{1}{12}\text{ h}$,

$$T_6 = \frac{1}{12}\left[\frac{24+60}{2} + 96 + 144 + 132 + 108 + 84\right] = \frac{606}{12} = 50.5\text{ kW h},$$
$$S_6 = \frac{1}{36}\left[24 + 60 + 4(96 + 132 + 84) + 2(144 + 108)\right] = \frac{1836}{36} = 51.0\text{ kW h}.$$

The charger's 150 kW sticker rating never decided the outcome; the area under the taper curve did. At \$0.40 per kilowatt-hour the session costs about \$20.40. The same integral runs continuously inside the battery-management system: integrating measured current over time, called coulomb counting, is how a phone, laptop, or electric vehicle knows its state of charge.

PUSH THE IDEA

Adaptive quadrature spends samples where curvature or estimated error is high. The same principle appears in rendering, finite-element meshes, and data acquisition: allocate computational effort where the function is hardest to represent.

COMPANION LAB: Simpson's Rule Studio

Choose a function or drag sample points. Compare rectangle, trapezoidal, and Simpson estimates against the exact area when available. Change interval count, add noise, and enable adaptive refinement to watch error evolve.

CHECK YOURSELF

1. Use the trapezoidal rule on $y = [2, 5, 10]$ with $h = 1$.
2. Use Simpson's rule on the same data.
3. Why must the composite Simpson rule use an even number of intervals?
4. Give two reasons a refined calculation may still disagree with physical area.

CHAPTER 17

Volume, Surface, and Material

Lab 17: Solid Shop

Original scope: Chapter XVII, printed pages 264–285

Prisms, cylinders, pyramids, cones, frusta, spheres, rings, flywheels, surface area, volume, and mass estimation.

THE 1899 IDEA

Castle develops solid mensuration as a direct bridge from geometry to material, capacity, and weight. Prisms and cylinders establish volume as base area times height. Pyramids and cones introduce a factor of one third; frusta and compound solids are built by decomposition.

Source: Castle (1899), pp. 264–285.

17.1 Volume is layered area

For any right prism of constant cross-sectional area A and length L,

$$V = AL. \tag{17.1}$$

Thus a cylinder has $V = \pi r^2 h$. A pyramid or cone with the same base and height has one third of the prism volume:

$$V_{\text{cone}} = \frac{1}{3}\pi r^2 h. \tag{17.2}$$

The dimensional signature is length cubed. Scaling all lengths by k multiplies volume and, for constant density, mass by k^3.

WORKED EXAMPLE: Filament for a 3D-printed bracket

A slicer estimates a bracket at 18.4 cm^3 of PLA, density 1.24 g/cm^3, printed from filament of diameter 1.75 mm, so $r = 0.0875$ cm.

$$m = 1.24(18.4) = 22.8\,\text{g},$$
$$A_{\text{filament}} = \pi(0.0875)^2 = 0.024{,}05\,\text{cm}^2,$$
$$L = \frac{18.4}{0.02405} = 765\,\text{cm} \approx 7.65\,\text{m}.$$

The filament is one long thin cylinder, so $V = AL$ runs in reverse to give length. A 1 kg spool holds about 335 m, and this part uses roughly 2% of it.

17.2 Hollow objects are differences

A tube of outer radius R, inner radius r, and length L has material volume

$$V = \pi(R^2 - r^2)L. \tag{17.3}$$

Thin-wall approximations replace $R^2 - r^2 = (R - r)(R + r)$ with approximately $2rt$ when thickness t is small compared with radius r. The approximation is useful only when its neglected terms are insignificant for the decision.

WORKED EXAMPLE: Mass of a steel tube

A tube is 1.20 m long, with outer diameter 60 mm and wall thickness 4 mm. Then $R = 0.030$ m and $r = 0.026$ m.

$$V = \pi(0.030^2 - 0.026^2)(1.20)$$
$$= 0.000{,}844\,\text{m}^3,$$
$$m = \rho V = 7850(0.000844) = 6.63\,\text{kg}.$$

The estimate excludes coatings, welds, and dimensional tolerances.

17.3 Surface and volume scale differently

Surface-to-volume ratio decreases as a similar object grows. This affects heat transfer, chemical reaction, structural weight, biological metabolism, and packaging. Doubling every length multiplies surface by four and volume by eight.

PUSH THE IDEA

Modern solid modeling calculates mass properties directly from a boundary or mesh, but validation still begins with simple bounding shapes. A computed volume should lie between plausible inner and outer bounds. If not, investigate units, wall direction, missing faces, or self-intersections.

COMPANION LAB: Solid Shop

Select a prism, tube, cone, frustum, sphere, or compound solid. Adjust dimensions, material density, and unit system. The lab reports volume, surface area, mass, cost, and surface-to-volume ratio, with an exploded decomposition of the formula.

CHECK YOURSELF

1. Find the volume of a cylinder with $r = 0.20$ m and $h = 1.5$ m.
2. Find the mass of that cylinder if density is 2700 kg/m^3.
3. A similar solid triples every dimension. By what factor do surface area and volume change?
4. Derive the hollow-cylinder formula by subtraction.

CHAPTER 18

Motion, Force, Work, and Power

Lab 18: Motion and Energy Bench

Original scope: Chapter XVIII, printed pages 286–298

Linear motion, momentum, force, mass, work, power, kinetic and potential energy, angular velocity, and curved motion.

THE 1899 IDEA

Castle connects kinematics to dynamics: motion describes change, force changes momentum, work transfers energy through distance, and power measures the rate of that transfer. His practical emphasis is consistent units and a clear distinction between mass and force.

Source: Castle (1899), pp. 286–298.

18.1 Motion is change with respect to time

Average velocity is displacement divided by elapsed time. Instantaneous velocity and acceleration are derivatives:

$$v = \frac{dx}{dt}, \qquad a = \frac{dv}{dt}. \tag{18.1}$$

For constant acceleration,

$$v = u + at, \qquad s = ut + \frac{1}{2}at^2, \qquad v^2 = u^2 + 2as. \tag{18.2}$$

These equations are not interchangeable recipes. Each assumes constant acceleration and relates a different set of known quantities.

18.2 Force and energy answer different questions

Newton's second law in its general form is $\mathbf{F} = d\mathbf{p}/dt$, where momentum $\mathbf{p} = m\mathbf{v}$. For constant mass, $\mathbf{F} = m\mathbf{a}$. Work by a constant aligned force is $W = Fs$; power is

$$P = \frac{dW}{dt} = Fv. \tag{18.3}$$

Kinetic and gravitational potential energy are

$$E_k = \frac{1}{2}mv^2, \qquad E_p = mgh. \tag{18.4}$$

Energy is scalar; momentum is vector. Both are conserved in appropriate closed systems, but they constrain interactions differently.

HISTORY THREAD: Emmy Noether: look for what does not change

Emmy Noether (1882–1935) reshaped modern algebra by emphasizing structures and the relationships that preserve them. In mathematical physics, Noether's theorem connects continuous symmetries with conservation laws: under the right conditions, invariance under shifts in time corresponds to conservation of energy. She produced this work while institutions repeatedly restricted her teaching because she was a woman; the Nazi government later dismissed her because she was Jewish.

Her method offers a survival-grade question for any complicated system: what can change, and what must remain invariant? Conservation laws turn that question into a calculation. Structure is not abstraction away from reality; sometimes it is the shortest route into it.

Historical note: MacTutor, "Emmy Noether".

WORKED EXAMPLE: Hoist power

A hoist raises 420 kg vertically by 7.5 m in 18 s. Ignoring losses,

$$W = mgh = 420(9.81)(7.5) = 30.9\,\text{kJ},$$
$$P = W/t = 30.9/18 = 1.72\,\text{kW}.$$

At 78% overall efficiency, required input power is $1.72/0.78 = 2.20$ kW.

18.3 Curved motion needs inward acceleration

Uniform circular motion has constant speed but changing velocity. Its inward acceleration is

$$a_c = \frac{v^2}{r} = r\omega^2. \tag{18.5}$$

The required centripetal force is not a new kind of force; it is the net inward force supplied by tension, friction, gravity, or another interaction.

18.4 Change written as an equation

The definitions $v = dx/dt$ and $a = dv/dt$ contain a larger idea. An equation that specifies a rate of change is a differential equation, and solving one means recovering the quantity from its law of change. The constant-acceleration formulas are exactly that: the solved form of $dv/dt = a$. Many working systems obey the first-order law

$$\frac{dT}{dt} = -k(T - T_{\text{amb}}), \tag{18.6}$$

which says temperature approaches ambient at a rate proportional to the remaining gap. Its solution is exponential decay of that gap:

$$T(t) = T_{\text{amb}} + (T_0 - T_{\text{amb}})\,e^{-kt}. \tag{18.7}$$

Differentiate the solution and the original law returns; that is the test of any proposed solution. The same equation, with different letters, discharges a capacitor, depressurizes a tank, and clears medication from a bloodstream.

WORKED EXAMPLE: How long until it is safe to touch?

A sealed motor controller shuts down at $T_0 = 82\,°\text{C}$ in a 22 °C room, and its logged cooling curve gives $k = 0.05/\text{min}$. When does the case reach 40 °C? Solve $40 = 22 + 60e^{-0.05t}$:

$$e^{-0.05t} = \frac{18}{60} = 0.30 \quad \Rightarrow \quad t = \frac{\ln(60/18)}{0.05} = \frac{1.204}{0.05} \approx 24\text{ min}.$$

When no clean solution exists, integrate the law numerically. Euler's method steps along the current slope, $T_{n+1} = T_n - hk(T_n - T_{\text{amb}})$: one 2 min step from 82 °C predicts $82 - 2(0.05)(60) = 76.0\,°\text{C}$ against the exact 76.3 °C, and shrinking h shrinks the error. A drone flight controller does precisely this with the equations of motion, hundreds of times per second, to hold its position.

PUSH THE IDEA

Energy accounting scales from a hoist to an electric vehicle or a data center. Define the system boundary, list energy stores and transfers, and separate peak power from total energy. A battery may hold enough energy for a task yet fail because it cannot deliver the required power safely.

COMPANION LAB: Motion and Energy Bench

Set mass, initial velocity, acceleration, distance, efficiency, and time. The lab animates motion and plots velocity, kinetic energy, work, and power. A circular mode links rpm, radius, tangential speed, and centripetal force.

CHECK YOURSELF

1. A vehicle accelerates from rest at 2.5 m/s² for 8 s. Find final speed and distance.
2. Find kinetic energy of 1200 kg moving at 15 m/s.
3. Find output power when 900 N acts at 3.2 m/s in the direction of motion.
4. Distinguish energy from power in one sentence.

CHAPTER 19

Vectors, Equilibrium, and Structures

Lab 19: Force Polygon

Original scope: Chapter XIX, printed pages 299–326

Vector specification, resultants, force polygons, equilibrium, moments, centroids, trusses, and moments of inertia.

THE 1899 IDEA

Castle represents a force graphically by a directed line whose length gives magnitude and whose arrow gives direction. Closed vector polygons express equilibrium. He extends the method to moments, parallel forces, centers of gravity, roof trusses, and area moments of inertia.

Source: Castle (1899), pp. 299–326.

19.1 Vectors carry magnitude and direction

A planar vector can be written from components:

$$\mathbf{F} = F_x\mathbf{i} + F_y\mathbf{j}, \qquad |\mathbf{F}| = \sqrt{F_x^2 + F_y^2}. \tag{19.1}$$

Vector addition is componentwise. Graphically, placing vectors head to tail constructs the same sum. A closed polygon means the resultant is zero.

19.2 Equilibrium requires force and moment balance

For a rigid body in static planar equilibrium,

$$\sum F_x = 0, \qquad \sum F_y = 0, \qquad \sum M_O = 0. \tag{19.2}$$

Moment about O is force times perpendicular lever arm. Choosing a moment center through unknown forces can eliminate them from an equation and simplify the solution.

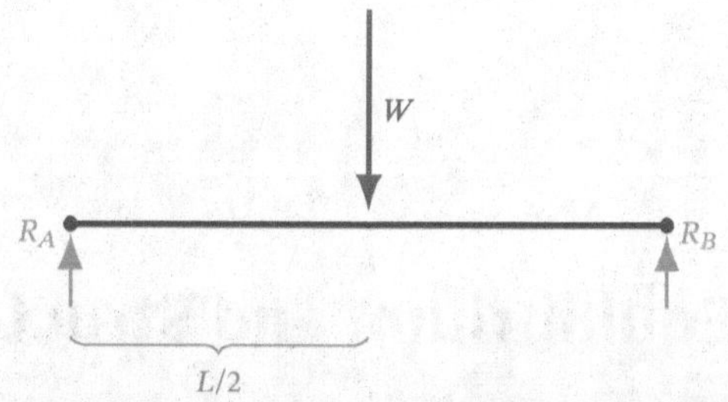

WORKED EXAMPLE: Simply supported beam

A 6 m beam carries a centered 12 kN load. Symmetry suggests equal reactions. Equilibrium confirms it:

$$\sum M_A = R_B(6) - 12(3) = 0 \Rightarrow R_B = 6\,\text{kN},$$
$$\sum F_y = R_A + R_B - 12 = 0 \Rightarrow R_A = 6\,\text{kN}.$$

The maximum bending moment occurs at midspan: $M_{max} = R_A(3) = 18\,\text{kN m}$.

19.3 Simultaneous equilibrium becomes matrix algebra

Each joint contributes force-balance equations, and the unknowns appear in several equations at once. Matrix notation organizes the whole system as a single object,

$$A\mathbf{x} = \mathbf{b}, \tag{19.3}$$

where A holds direction cosines fixed by the geometry, $\mathbf{x}$ lists the unknown forces, and $\mathbf{b}$ lists the applied loads. For two unknowns the inverse is explicit:

$$\begin{bmatrix} a & b \\ c & d \end{bmatrix}^{-1} = \frac{1}{ad - bc} \begin{bmatrix} d & -b \\ -c & a \end{bmatrix}, \qquad ad - bc \neq 0. \tag{19.4}$$

A zero determinant is not a bookkeeping failure. It means the geometry cannot resist the load: the structure is a mechanism.

WORKED EXAMPLE: Two cables solved by matrix inverse

A 900 N camera rig hangs from two cables anchored to opposite walls, one at 30° and one at 45° above horizontal. Joint equilibrium in matrix form is

$$\begin{bmatrix} \cos 30^\circ & -\cos 45^\circ \\ \sin 30^\circ & \sin 45^\circ \end{bmatrix} \begin{bmatrix} T_1 \\ T_2 \end{bmatrix} = \begin{bmatrix} 0 \\ 900 \end{bmatrix}.$$

The determinant is $\cos 30^\circ \sin 45^\circ + \cos 45^\circ \sin 30^\circ = \sin 75^\circ = 0.966$, so the inverse exists. Applying it,

$$T_1 = \frac{900\cos 45^\circ}{\sin 75^\circ} = 659\text{ N},$$
$$T_2 = \frac{900\cos 30^\circ}{\sin 75^\circ} = 807\text{ N}.$$

The steeper cable carries more of the load, as intuition expects. Check vertically: $T_1 \sin 30^\circ + T_2 \sin 45^\circ = 329.4 + 570.6 = 900$ N.

The same algebra now runs far beyond structures. A computer rotates an image by multiplying every point by $R = \begin{bmatrix} \cos\theta & -\sin\theta \\ \sin\theta & \cos\theta \end{bmatrix}$; a robot arm chains such matrices to place its gripper; least-squares fitting solves $A^{\mathsf{T}}A\,\mathbf{x} = A^{\mathsf{T}}\mathbf{b}$; and graphics processors earned a second career training neural networks because rendering and learning are both, at bottom, matrix multiplication.

19.4 Centroid and moment of inertia describe distribution

The centroid locates the first moment balance of an area. The second moment of area

$$I = \int y^2\,dA \tag{19.5}$$

measures how area is distributed from an axis and strongly affects bending stiffness. It is not the same as mass moment of inertia, though the mathematical form is related.

PUSH THE IDEA

Finite-element analysis replaces a continuous structure with many connected elements and solves a large equilibrium system. A colorful stress plot does not remove the need for free-body diagrams, reactions, boundary-condition checks, or hand-calculated bounds.

COMPANION LAB: Force Polygon

Add forces by magnitude and angle, inspect components and the resultant, then close the polygon for equilibrium. Beam mode calculates support reactions, shear, and bending moment; truss mode solves joint forces and labels tension or compression.

CHECK YOURSELF

1. Resolve 500 N at 35° above horizontal into components.
2. Find the resultant of $(3, 4)$ kN and $(-1, 2)$ kN.
3. A 10 m simply supported beam carries 20 kN at midspan. Find reactions.
4. Explain why moving area away from a neutral axis increases second moment of area.

CHAPTER 20

Curves, Machines, Stress, and Strain

Lab 20: Curve and Material Lab

Original scope: Chapter XX, printed pages 327–353

Squared paper, plotting, machine relationships, maxima and minima, slope, variable work, springs, stress, strain, and elastic modulus.

THE 1899 IDEA

Castle's broad final applications chapter uses squared paper as an experimental workspace. Curves reveal machine laws, maxima and minima, spring behavior, stress-strain relationships, and work by variable force. The graph becomes a bridge between observation and equation.

Source: Castle (1899), pp. 327–353.

20.1 Slope and area are local and accumulated change

The slope of $y = f(x)$ is the derivative dy/dx. The area under a force-distance curve is work:

$$W = \int_{x_1}^{x_2} F(x)\,dx. \tag{20.1}$$

These are inverse viewpoints. Differentiation asks how rapidly a quantity is changing at a point. Integration asks how those changes or contributions accumulate across an interval.

20.2 Stress separates load from size

Normal engineering stress is force divided by original cross-sectional area; engineering strain is extension divided by original length:

$$\sigma = \frac{F}{A}, \qquad \varepsilon = \frac{\Delta L}{L_0}. \tag{20.2}$$

In the linear elastic region, Hooke's law is

$$\sigma = E\varepsilon, \tag{20.3}$$

where E is Young's modulus. Stress has pressure units; strain is dimensionless.

HISTORY THREAD: Euler predicts a sudden failure

Leonhard Euler applied analysis to the bending and stability of slender columns. The ideal Euler buckling load scales with stiffness and falls with the square of effective length. That square is the practical warning: doubling a column's unsupported length can reduce its ideal buckling load to one quarter.

Buckling also changes the philosophy of safety. A member can fail through loss of stability before its material reaches a simple crushing strength. Reality may be governed not by the largest number on a material table, but by geometry, boundary conditions, and imperfections. Applied mathematics is powerful when it reveals the failure mode we were not yet watching.

Historical note: MacTutor, "Leonhard Euler"; see also standard treatments of Euler buckling.

WORKED EXAMPLE: Extension of a steel tie

A steel tie is 2.0 m long, has area 300 mm^2, carries 45 kN, and has $E = 200$ GPa.

$$\sigma = \frac{45000}{300 \times 10^{-6}} = 150\,\text{MPa},$$
$$\varepsilon = \frac{150 \times 10^{6}}{200 \times 10^{9}} = 0.00075,$$
$$\Delta L = \varepsilon L_0 = 0.00075(2.0) = 1.5\,\text{mm}.$$

The calculation assumes axial load, uniform area, linear elasticity, and no stress concentration.

20.3 Springs store area under a line

For a linear spring $F = kx$, work stored is

$$W = \int_0^x kx\,dx = \frac{1}{2}kx^2. \tag{20.4}$$

The triangular area beneath the force-extension line is visible on the graph. If loading and unloading curves differ, the enclosed loop represents energy dissipated as hysteresis.

WORKED EXAMPLE: Energy stored in a bike-shock spring

A mountain-bike coil shock has spring rate $k = 60$ N/mm (6.0×10^4 N/m) and compresses $x = 55$ mm at full travel.

$$F_{\max} = kx = 6.0 \times 10^4(0.055) = 3300\,\text{N},$$
$$W = \frac{1}{2}kx^2 = \frac{1}{2}(6.0 \times 10^4)(0.055)^2 = 90.8\,\text{J}.$$

The triangular area under the force-extension line gives the same result, $\frac{1}{2}(3300)(0.055) = 90.8\,\mathrm{J}$, roughly an 80 kg rider dropping 12 cm; the damper and tires absorb the rest of a real landing.

PUSH THE IDEA

Data-driven material models fit nonlinear, rate-dependent, or temperature- dependent behavior. Model selection should follow the operating range. A line fit to the elastic region should not be extrapolated through yield and fracture. The plot is an argument about domain, not merely a picture.

COMPANION LAB: Curve and Material Lab

Plot force-extension or stress-strain data, choose a fit region, compute slope and area, and identify elastic limit, yield, and energy absorption. Machine mode plots input-output laws and searches for maxima, minima, and efficiency peaks.

CHECK YOURSELF

1. Find stress for 24 kN on 160 mm^2.
2. Find strain for 0.8 mm extension over 500 mm.
3. A 300 N/m spring extends 0.12 m. Find stored energy.
4. State two assumptions behind the steel-tie worked example.

CHAPTER 21

Formulae as Executable Models

Lab 21: Formula Runner

Original scope: Chapter XXI, printed pages 354–370

Evaluation of algebraic and trigonometric formulas, substitution, units, order of operations, validation, and translation into code.

THE 1899 IDEA

Castle closes the instructional chapters with sustained formula evaluation. The student must identify variables, substitute consistently, preserve signs and grouping, use tables correctly, and judge whether the numerical result is reasonable. It is a capstone in disciplined execution.

Source: Castle (1899), pp. 354–370.

21.1 A formula is a compact model

A formula names a relationship between quantities. Evaluation should follow a repeatable sequence:

1. state the meaning and unit of every variable;
2. convert inputs to a compatible unit system;
3. write the substituted expression before calculating;
4. preserve grouping and calculator angle mode;
5. estimate scale and sign;
6. calculate with guard digits;
7. report with units and justified precision;
8. substitute or compare against a limiting case.

This sequence is longer than typing numbers, and much shorter than diagnosing an untraceable result.

HISTORY THREAD: Euler and the compression of thought

Euler's achievement was not only the huge range of problems he solved. He also helped stabilize notation now woven into ordinary mathematics, including $f(x)$ for a function and widespread uses of e, i, π, and summation notation. No symbol was created in a vacuum, and not every convention was his alone, but Euler's writing helped make a coherent working language portable.

Notation is cognitive equipment. A good symbol does not merely shorten a sentence; it makes transformations visible and frees attention for structure. The danger is forgetting what the compressed marks mean. Formula literacy therefore requires both speeds: fluency with the shorthand and the ability to expand it back into quantities, assumptions, and consequences.

Historical note: MacTutor, "Leonhard Euler".

21.2 Dimensions catch structural mistakes

If a formula claims distance but its right-hand side has units of speed, a time factor is missing. Dimensionally consistent equations are invariant under unit conversion. Dimensional consistency is necessary, not sufficient: both $s = ut + at^2$ and $s = ut + \frac{1}{2}at^2$ have distance dimensions, but only one has the correct coefficient for constant acceleration from the stated initial condition.

WORKED EXAMPLE: Evaluate and validate a discharge formula

For ideal efflux, $v = \sqrt{2gh}$. With $g = 9.81\,\mathrm{m/s^2}$ and $h = 1.80\,\mathrm{m}$,

$$v = \sqrt{2(9.81)(1.80)} = 5.94\,\mathrm{m/s}.$$

Dimensional check: $\sqrt{(\mathrm{m\,s^{-2}})(\mathrm{m})} = \mathrm{m\,s^{-1}}$. Limiting check: as $h \to 0$, $v \to 0$. Scaling check: quadrupling h doubles v, because height is under a square root.

21.3 Translate meaning, not punctuation, into code

```
from math import sqrt

def ideal_efflux_speed(height_m, gravity=9.81):
    if height_m < 0:
        raise ValueError("height must be nonnegative")
    return sqrt(2 * gravity * height_m)
```

The code adds a domain check absent from the bare formula. A production version would also document units, test known cases, and perhaps use a quantity library that prevents incompatible-unit operations.

WORKED EXAMPLE: Runtime of a power bank

A power bank stores 20,000 mA h at a nominal 3.7 V and feeds a laptop drawing 30 W through a converter of 90% efficiency.

$$E = \frac{20000(3.7)}{1000} = 74.0\,\text{W h},$$
$$t = \frac{0.90(74.0)}{30} = 2.22\,\text{h} \approx 2\,\text{h}\ 13\,\text{min}.$$

Dimensional check: W h/W = h. Limiting check: doubling the load halves the runtime.

```
def
    if            or
        raise             "load and efficiency must be positive"
    return
```

The function reproduces 2.22 h and refuses a zero or negative load before the bare formula can divide by it.

PUSH THE IDEA

A formula can become a function, spreadsheet cell, simulation component, or real-time controller. Each transition adds failure modes: stale references, unit mismatches, sampling effects, overflow, invalid domains, and hidden defaults. The remedy is the same discipline Castle taught: state, substitute, calculate, and verify - now supported by tests and type systems.

COMPANION LAB: Formula Runner

Select any formula used in the book, inspect variable meanings and units, and evaluate it from editable inputs. The runner shows symbolic substitution, dimensional checks, sensitivity, limiting cases, and a short generated code equivalent. Saved examples remain on the reader's device.

CHECK YOURSELF

1. Evaluate $v = \sqrt{2gh}$ for $h = 5.0$ m.
2. Check dimensions of $E = \frac{1}{2}mv^2$.
3. State the limiting behavior of $A = \pi r^2$ as $r \to 0$ and as r doubles.
4. List three validations to perform when translating a formula into code.

Answer Key

Answers are concise so that the reader can check method and scale. Equivalent forms and appropriately rounded results are acceptable.

Chapter 1

1. 1143 mm.
2. Approximately 24.61 mm to 24.63 mm.
3. $5/1023 = 0.004{,}89$ V per count.
4. Repeated readings may agree closely while a zero or calibration bias moves all of them away from the true value.

Chapter 2

1. 4500 cm^2.
2. 2.70 kg/L=2700 kg/m^3.
3. $720/1000 = 0.72$, or 72% submerged.
4. The hull encloses air, so the ship's average density over its displaced volume can be less than water's density.

Chapter 3

1. $5\pi/4$ rad.
2. $s = r\theta = 0.48$ m.
3. $\omega = 2\pi(900)/60 = 94.25$ rad/s.
4. The nearly fixed start/stop reaction error is distributed across 20 cycles, reducing its relative effect on one period.

Chapter 4

1. $61/60 = 1\frac{1}{60}$.
2. Greatest common factor 42; least common multiple 252.
3. $(40 \times 12 + 15 \times 18)/55 = \$13.64$ per unit.
4. For skewed quantities such as household income, the median better represents the middle observation because extremes pull the mean.

Chapter 5

1. 4 m.
2. 15 cm.
3. $(966 - 840)/840 = 15\%$.
4. $1.20(0.80) = 0.96$, a net decrease of 4%.

Chapter 6

1. $x = 11$.
2. $3x(2x - 5)$.
3. a^5/b^3.
4. $a = 2(s - ut)/t^2$.

Chapter 7

1. 18.15 cm$\leq L <$18.25 cm.
2. $|49.8 - 50.0|/50.0 = 0.4\%$.
3. 0.00988.
4. Worst-case relative uncertainty is approximately $\Delta a/a + \Delta b/b + \Delta c/c$.

Chapter 8

1. $x = 6$.
2. $x = 5, y = 2$.
3. No solution; the left sides are proportional but the constants are inconsistent.
4. $x + y = 15$ and $8x + 13y = 155$, with $x, y \geq 0$ integers.

Chapter 9

1. $m = (20 - 5)/(8 - 2) = 2.5$.
2. $y = 24$.
3. 15 m.
4. The relationship may curve, saturate, or change mechanism outside the observed range.

Chapter 10

1. 5 and −3.
2. $x = \ln 50/\ln 3 \approx 3.561$.
3. $2\ln a + \frac{1}{2}\ln b - \ln c$.
4. No real logarithm exists there, and ratios to zero are undefined.

Chapter 11

1. $18 \sin 24° = 7.32$ m.
2. $\tan^{-1}(0.8/9.5) = 4.81°$.
3. $60 \tan 18° + 1.5 \approx 21.0$ m.
4. Degree inputs interpreted as radians, or the reverse, represent entirely different angles.

Chapter 12

1. About 0.51; the exact product is 0.5146.
2. $(4.7 \times 10^3)/(2.3 \times 10^{-2}) \approx 2.04 \times 10^5$.
3. Scale length and reading resolution limit significant figures even when the operator interpolates.
4. Hidden steps increase speed but can conceal a wrong assumption, stale input, or unit mismatch.

Chapter 13

1. 83.
2. $3(2^{12} - 1) = 12{,}285$.
3. $5000(1.04)^8 = \$6{,}842.85$.
4. $10000/(1.06)^5 = \$7{,}472.58$.

Chapter 14

1. $\sqrt{4^2 + 3^2 + 6^2} = \sqrt{61} \approx 7.81$.
2. The z coordinate.
3. When the line is parallel to the projection plane.
4. Orthographic projection preserves parallel lines and scale with depth; perspective projection makes distant objects appear smaller.

Chapter 15

1. $59.5\ m^2$.
2. $\pi(90^2 - 65^2) \approx 12{,}174\ mm^2$.
3. Perimeter by 2; area by 4.
4. Absolute errors in the large areas can be comparable with the small difference, producing large relative uncertainty.

Chapter 16

1. $T = \frac{1}{2}(2 + 10) + 5 = 11$.

2. $S = (2 + 4(5) + 10)/3 = 32/3 \approx 10.667$.
3. Each quadratic interpolant spans two adjacent intervals.
4. Examples include measurement noise, missing boundary detail, wrong spacing, wrong sample order, or a scale error.

Chapter 17

1. $\pi(0.20)^2(1.5) = 0.1885\,\mathrm{m}^3$.
2. Approximately 509 kg.
3. Surface area by 9; volume by 27.
4. Subtract the inner cylinder volume $\pi r^2 L$ from the outer volume $\pi R^2 L$.

Chapter 18

1. $v = at = 20\,\mathrm{m/s}$ and $s = \frac{1}{2}at^2 = 80\,\mathrm{m}$.
2. 135 kJ.
3. 2880 W=2.88 kW.
4. Energy is capacity or quantity of transferred work; power is the rate at which energy is transferred.

Chapter 19

1. $F_x = 409.6\,\mathrm{N}, F_y = 286.8\,\mathrm{N}$.
2. $(2, 6)$ kN, magnitude $\sqrt{40} = 6.32\,\mathrm{kN}$.
3. 10 kN upward at each support.
4. The integrand $y^2 dA$ gives area farther from the axis greater weight.

Chapter 20

1. 150 MPa.
2. $0.8/500 = 0.0016$.
3. $\frac{1}{2}(300)(0.12)^2 = 2.16\,\mathrm{J}$.
4. Any two: axial loading, uniform cross-section, linear elasticity, homogeneous material, negligible stress concentration.

Chapter 21

1. $\sqrt{2(9.81)(5.0)} = 9.90\,\mathrm{m/s}$.
2. $\mathrm{kg}(\mathrm{m\,s^{-1}})^2 = \mathrm{kg\,m^2\,s^{-2}} = \mathrm{J}$.
3. $A \to 0$ as $r \to 0$; doubling r multiplies A by 4.

4. Examples: unit check, domain check, known-case test, limiting-case test, inverse substitution, and independent numerical comparison.

Glossary

Absolute error
Difference between an approximate or measured value and a reference value, expressed in the same unit.

Acceleration
Rate of change of velocity with time.

Accuracy
Closeness of a result to the true or accepted value.

Algorithm
A defined sequence of operations for producing a result.

Arithmetic progression
Sequence with a constant difference between successive terms.

Buoyancy
Upward fluid force equal to the weight of displaced fluid.

Centroid
Geometric center defined by first moments of area.

Compound interest
Growth in which each period's change applies to the accumulated balance.

Density
Mass per unit volume.

Derivative
Instantaneous rate of change of a quantity; the slope of its graph at a point.

Differential equation
Equation that specifies a quantity through its rate of change; solving it recovers the quantity itself.

Dimension
Physical kind of a quantity, such as length, mass, or time.

Equilibrium
State in which resultant force and resultant moment are zero.

Euler's method
Numerical solution of a differential equation by repeated short steps along the current slope.

Extrapolation
Estimation outside the observed or tabulated range.

Geometric progression
Sequence with a constant ratio between successive terms.

Interpolation
Estimation between known observations.

Logarithm
Exponent required to obtain a number from a stated base.

Longitude
Angular position east or west of a reference meridian; historically found by comparing local solar time with carried reference time.

Matrix
Rectangular array of numbers that represents a linear system or transformation as a single object.

Measurement
Comparison of a quantity with a defined unit.

Mil (angular)
Angle unit near one milliradian; 6400 mils span a circle, and one mil subtends about one metre at one kilometre.

Moment
Turning effect of a force about a point or axis.

Momentum
Product of mass and velocity; a vector quantity.

Numerical integration
Approximation of accumulated area or contribution from sampled values.

Orthographic projection
Parallel projection onto a plane, commonly used for technical views.

Power
Rate of energy transfer or work.

Precision
Closeness of repeated results to one another.

Proportion
Equality of two ratios.

Radian
Angle subtending an arc equal in length to the radius.

Ratio
Dimensionless comparison of like quantities expressed in common units.

Relative error
Absolute error divided by a reference magnitude.

Resolution
Smallest increment that a measuring system can distinguish or report.

Scale
Mapping between represented and physical magnitude.

Significant figures
Digits used to communicate the precision of a value.

Simpson's rule
Numerical integration rule based on quadratic interpolation over pairs of intervals.

Slope
Ratio of vertical change to horizontal change; derivative for a smooth curve.

Strain
Relative deformation, commonly extension divided by original length.

Stress
Internal force intensity, commonly force divided by cross-sectional area.

Trapezoidal rule
Numerical integration rule based on linear interpolation between samples.

Uncertainty
Quantified doubt associated with a measured or inferred value.

Unit
Agreed reference quantity used in measurement.

Vector
Quantity having magnitude and direction.

Work
Energy transferred when force acts through displacement.

Source Map

The table maps this edition to the printed pagination of Castle's 1899 first edition and to the page numbers inside the University of Michigan Google Books scan. PDF page number equals printed page number plus 17 throughout the body.

Ch.	Original scope	Print	PDF
1	Measurement and units of length	1–6	18–23
2	Area, volume, density, Archimedes	7–21	24–38
3	Time and angular measurement	22–38	39–55
4	Arithmetic, fractions, decimals, averages	39–58	56–75
5	Powers, roots, ratio, proportion, percentage	59–75	76–92
6	Algebraic operations, factors, indices	76–95	93–112
7	Approximation, measures, multiples, fractions	96–109	113–126
8	Simple and simultaneous equations	110–127	127–144
9	Graphic methods, scales, right triangles	128–141	145–158
10	Logarithms	142–157	159–174
11	Trigonometric functions and indirect measurement	158–177	175–194
12	The slide rule	178–188	195–205
13	Progressions, interest, discount	189–199	206–216
14	Orthographic projection	200–227	217–244
15	Plane mensuration	228–250	245–267
16	Trapezoidal rule, Simpson's rule, planimeter	251–263	268–280
17	Solid mensuration	264–285	281–302
18	Motion, force, work, power, energy	286–298	303–315
19	Vectors, equilibrium, structures, inertia	299–326	316–343
20	Curves, machines, stress, strain	327–353	344–370
21	Evaluation of formulae	354–370	371–387

Editorial OCR files are stored alongside the project source as `source/extracted/chapter-01.txt` through `source/extracted/chapter-21.txt`. They preserve scan and printed-page markers for audit but are not treated as publication-ready prose.

The World Behind the Formulas

Mathematics has no single birthplace. It grew wherever people counted goods, measured land, tracked the sky, designed structures, shared inheritances, or asked what patterns must be true. The history threads in this edition are deliberately brief. They are doorways, not a claim to name every culture or every contributor.

The threads also resist a tidy parade of isolated geniuses. Knowledge moved through translation, travel, teaching, argument, and rediscovery. A theorem can carry one person's name while preserving work from many communities. A symbol can be invented in one place, refined in another, and become ordinary everywhere.

Measurement and administration - Egypt and Mesopotamia

Chapter 1 begins with standards, trade, land, and the practical origins of recorded mathematics.

Place value and calendars - Mesopotamia and the Maya

Chapter 3 compares two powerful number systems shaped by timekeeping and astronomy.

Algorithms - China

Chapter 4 follows counting rods, signed numbers, and systematic elimination.

Algebra - the Arabic-language scholarly world

Chapter 6 visits Baghdad and al-Khwarizmi, while recognizing the multilingual, multi-faith network through which the subject developed.

Insight and proof - Ramanujan and India

Chapter 7 asks what happens when intuition runs ahead of formal demonstration.

Zero and negative numbers - India

Chapter 8 considers Brahmagupta's attempt to make a coherent arithmetic of zero, fortunes, and debts.

Coordinates - early-modern Europe

Chapter 9 follows the joining of algebra and geometry associated with Descartes, while noting that ideas mature through schools and predecessors.

Trigonometry - India and the Islamic world

Chapter 11 traces a chain from Indian sine tables to al-Tusi's independent discipline of plane and spherical trigonometry.

Right triangles - Mesopotamia, India, China, and Greece

Chapter 15 puts the Pythagorean theorem back into its cross-cultural history.

Infinite processes - China and Kerala

Chapter 16 pairs Liu Hui's refinement of polygons with Madhava's infinite series: two ways finite steps can approach an exact result.

Symmetry and structure - Emmy Noether

Chapter 18 shows how a barred academic outsider changed algebra and physics by looking for what remains invariant.

Notation and application - Leonhard Euler

Chapters 20 and 21 distinguish Euler's engineering mathematics from his gift for notation that makes complicated thought easier to execute.

A formula has a history, but truth is not owned by its label.

Names help us remember. Networks help us understand. The practical reader should learn both: who left a visible mark, and which older or less visible hands made that mark possible.

Sources and Further Reading

Castle, Frank.
Elementary Practical Mathematics. London: Macmillan and Company, Limited, 1899. University of Michigan scan. Google Books volume `Abk2AAAAMAAJ`. This is the public-domain base text for the edition.

Bureau International des Poids et Mesures.
The International System of Units (SI Brochure). Current definitions, symbols, and conventions for SI measurement.

National Institute of Standards and Technology.
Guidelines for Evaluating and Expressing the Uncertainty of NIST Measurement Results. A practical bridge from measured values to uncertainty statements.

OpenStax.
Precalculus and *University Physics*. Openly licensed modern references for algebra, trigonometry, vectors, mechanics, and energy.

MacTutor History of Mathematics Archive.
University of St Andrews. Biographies and topic essays used for the historical threads include measurement, Egyptian and Babylonian mathematics, Maya mathematics, Liu Hui and the *Nine Chapters*, Brahmagupta and zero, al-Khwarizmi and Arabic-language mathematics, Ramanujan, Descartes, Nasir al-Din al-Tusi, Pythagorean traditions, Madhava, Emmy Noether, and Euler. `https://mathshistory.st-andrews.ac.uk/`

Publisher information.
4DAT title and edition updates are posted at `https://www.iamfrankcastle.com`.

Editorial principle

Historical text is most useful when it remains inspectable. Every modernization in this book should be reversible: the reader can locate the original scope, see the updated notation, reproduce the example, and test the extension in the companion lab. That chain of evidence is the book's central method.

Index

www.ingramcontent.com/pod-product-compliance
Lightning Source LLC
LaVergne TN
LVHW030727061126
851857LV00019B/2029
9781972327180